Quantitative Methods in the Humanities and Social Sciences

Quantitative Methods in the Humanities and Social Sciences is a book series designed to foster research-based conversation with all parts of the university campus – from buildings of ivy-covered stone to technologically savvy walls of glass. Scholarship from international researchers and the esteemed editorial board represents the far-reaching applications of computational analysis, statistical models, computer-based programs, and other quantitative methods. Methods are integrated in a dialogue that is sensitive to the broader context of humanistic study and social science research. Scholars, including among others historians, archaeologists, new media specialists, classicists and linguists, promote this interdisciplinary approach. These texts teach new methodological approaches for contemporary research. Each volume exposes readers to a particular research method. Researchers and students then benefit from exposure to subtleties of the larger project or corpus of work in which the quantitative methods come to fruition.

More information about this series at http://www.springer.com/series/11748

Arjuna Tuzzi
Editor

Tracing the Life Cycle of Ideas in the Humanities and Social Sciences

Springer

Editor
Arjuna Tuzzi
Department of Philosophy, Sociology, Education
and Applied Psychology
University of Padova
Padova, Italy

ISSN 2199-0956 ISSN 2199-0964 (electronic)
Quantitative Methods in the Humanities and Social Sciences
ISBN 978-3-030-07296-4 ISBN 978-3-319-97064-6 (eBook)
https://doi.org/10.1007/978-3-319-97064-6

Softcover re-print of the Hardcover 1st edition 2018

This Springer imprint is published by the registered company Springer Nature Switzerland AG
The registered company address is: Gewerbestrasse 11, 6330 Cham, Switzerland

Foreword

Some years ago, I made, together with my students, some experiments aimed to test the Piotrowski-Altmann law on textual data from newspapers. The Piotrowski-Altmann law explains and describes the dynamics of the spread of new elements in a language and the dynamics of how elements of a language disappear. The formula which represents this law is

$$p(t) = \frac{1}{1 + ae^{-bt}}$$

It can be obtained as the solution to a differential equation which describes the dynamics of language change as a function of time. Apparently, the parameter b represents the velocity of change and can be interpreted as a bunch of linguistic and extralinguistic factors. The results of these tests gave perfect support to the hypothesis on language change and showed various forms of temporal behaviour of the function. Some words were on the increase; others could be observed while they were losing momentum. A special group reached a peak within 1 day and started decreasing the next day. Of course, there was no hope to single out the individual factors which contributed to the empirical values of the parameters and thus to a detailed interpretation of our results. We were happy enough with the empirical support to the law and a catalogue of several progression forms we found and could interpret in individual cases.

When Arjuna Tuzzi told me that she was planning a project based on distant reading using a quantitative approach aimed at data on the "history of ideas" in several scientific disciplines, I was not very optimistic at a first thought. It was clear that the search of such a history of concepts was methodologically very similar to the dynamics of linguistic elements because the concepts, or ideas, as taken from texts, are found in the form of terms in texts. I remembered my impressions from the experiments with my students. The results were excellent from a pure scientific point of view but did not look useful with respect to a chance to apply them. But then I thought: "What about if someone smarter than I am turned the process the other way round? Starting from one or two extra-linguistic factors and analysing the frequency dynamics of words or chunks found in the texts?". This was exactly the idea behind Arjuna Tuzzi's plan. And now I became enthusiastic.

A member of the scientific community has always some knowledge about his/her discipline: there are concepts, research questions, pioneers and important personalities, significant publications, debates and controversies, leading paradigms, failures and many more, which an informed colleague will be familiar with. On the other hand, no one is able to cover a discipline totally. The older a discipline, the harder a good picture on the basis of individual descriptions will be. After some decades, even a relatively young science becomes not even remotely comprehensible by a single person. Young colleagues are not yet able to gain an overview; older ones are less open to new developments. Thus, personal knowledge of a discipline is always incomplete and biased. A more complete picture can be obtained, of course, by reading as many relevant original books and articles as possible. This would become a project for decades, while the corresponding discipline keeps changing. Such a situation calls for statistics—the only method to collect reliable information in spite of fragmentary data. The project Arjuna Tuzzi was talking about suddenly seemed to provide the only possible way to achieve a "history of ideas" in several disciplines from texts and other data sources.

Now, I am tracking the project with rapt attention.

University of Trier
Trier, Germany

Reinhard Köhler

Contents

1 **Introduction: Tracing the History of a Discipline Through Quantitative and Qualitative Analyses of Scientific Literature** 1
Arjuna Tuzzi

Part I Tracing the Life-Cycle of Ideas

2 **Tracing the Words of the Analytic Turn in the Journal of Philosophy** 25
Giuseppe Spolaore and Pierdaniele Giaretta

3 **Exploring the History of American Sociology Through Topic Modelling** 45
Giuseppe Giordan, Chantal Saint-Blancat, and Stefano Sbalchiero

4 **Histories of Social Psychology in Europe and North America, as Seen from Research Topics in Two Key Journals** 65
Valentina Rizzoli

5 **First Steps in Shaping the History of Linguistics in Italy: The Archivio Glottologico Italiano** 87
Giovanni Urraci and Michele A. Cortelazzo

6 **The Recent History of Statistics: Comparing Temporal Patterns of Word Clusters** 105
Matilde Trevisani and Arjuna Tuzzi

Part II Concepts and Methods

7 **Treat Texts as Data but Remember They Are Made of Words: Compiling and Pre-processing Corpora** 133
Stefano Ondelli

8 **Automatic Multiword Identification in a Specialist Corpus** 151
Pasquale Pavone

9 **Functional Data Analysis and Knowledge-Based Systems** 167
Matilde Trevisani

10 **Topic Detection: A Statistical Model and a Quali-Quantitative Method** 189
Stefano Sbalchiero

11 **What Have We Learnt? Some Concluding Remarks** 211
Arjuna Tuzzi

Contributors

Michele A. Cortelazzo University of Padova, Padova, Italy

Pierdaniele Giaretta University of Padova, Padova, Italy

Giuseppe Giordan University of Padova, Padova, Italy

Stefano Ondelli University of Trieste, Trieste, Italy

Pasquale Pavone Università degli Studi di Modena e Reggio Emilia, Modena, Italy

Valentina Rizzoli University of Padova, Padova, Italy

Chantal Saint-Blancat University of Padova, Padova, Italy

Stefano Sbalchiero University of Padova, Padova, Italy

Giuseppe Spolaore University of Padova, Padova, Italy

Matilde Trevisani University of Trieste, Trieste, Italy

Arjuna Tuzzi Department of Philosophy, Sociology, Education and Applied Psychology, University of Padova, Padova, Italy

Giovanni Urraci University "Ca' Foscari" Venice, Venice, Italy

Abbreviations

AJS	American Journal of Sociology
ASA	American Sociological Association
ASR	American Sociological Review
ATD	Analysis of textual data
CA	Correspondence analysis
CC	Curve clustering
ECU	Elementary context units
ETD	Emerging topic detection
EASP	European Association of Social Psychology
EJSP	European Journal of Social Psychology
EDA	Exploratory data analysis
ETD	Emerging topic detection
FD	Functional data
FDA	Functional data analysis
FPCA	Functional principal component analysis
GCV	Generalized cross-validation
HDP	Hierarchical Dirichlet process
IE	Information extraction
IR	Information retrieval
JASA	Journal of the American Statistical Association
JPSP	Journal of Personality and Social Psychology
KWIC	Keyword in Context
KBS	Knowledge-based system
LNRE	Large number of rare events
LDA	Latent Dirichlet allocation
LSI	Latent semantic indexing
ML	Machine learning
MWE	Multiword expression
MI	Mutual information
NLP	Natural language processing
POS	Part-of-speech

PLSA	Probabilistic latent semantic analysis
PASA	Publications of the American Statistical Association
QASA	Quarterly Publications of the American Statistical Association
RE	Regular expression
RMS	Root mean square
SVD	Singular value decomposition
TM	Text mining
TDT	Topic Detection and Tracking
WSD	Word sense disambiguation

Chapter 1
Introduction: Tracing the History of a Discipline Through Quantitative and Qualitative Analyses of Scientific Literature

Arjuna Tuzzi

Contents

1.1 Quantitative Methods and History of Ideas 2
1.2 Tracks on the Ground: What Methods for What Purposes 4
1.3 Tracing the History of Words: A Quantitative Way 6
1.4 Objectives and Procedures 7
  1.4.1 Selection of Journals and Corpus Description 7
  1.4.2 Correspondence Analysis (CA) 9
  1.4.3 Identification of Keywords 10
  1.4.4 Curve Clustering 10
  1.4.5 Topic Detection 11
1.5 Chapters Outline 12
1.6 About This Book 12
Appendix 13
  A Brief Overview on Correspondence Analysis 13
  An Example 16
References 19

Abstract The chapters of this book are concerned with learning of the evolution of ideas (theories, concepts, methods, and application domains) and of the history of a discipline, by means of the temporal evolution of word occurrences in papers published by scientific journals. The work carried out for each of the areas involved in the project (philosophy, sociology, psychology, linguistics, statistics) pursued different objectives: to obtain a first overview of the relationship between time and contents in order to observe latent temporal patterns; to identify relevant keywords; to cluster keywords portraying similar temporal patterns; to identify latent dynamics of cluster keywords; and to identify relevant topics as groups of related words. The

A. Tuzzi (✉)
Department of Philosophy, Sociology, Education and Applied Psychology, University of Padova, Padova, Italy
e-mail: arjuna.tuzzi@unipd.it

A. Tuzzi (ed.), *Tracing the Life Cycle of Ideas in the Humanities and Social Sciences*, Quantitative Methods in the Humanities and Social Sciences,
https://doi.org/10.1007/978-3-319-97064-6_1

contributions identified and analysed the main subject matters that, at the time of publication, were considered relevant by mainstream journals and offer new viewpoints to read and understand the evolution of a discipline. The interdisciplinary debate triggered by this research work is innovative because quantitative methods for text analysis have been used in areas of human and social sciences, which are traditionally studied through qualitative approaches, and also represents a positive experience since new paths have been explored by pooling together the qualitative and quantitative research methods, traditions, and expertise of different disciplines.

Keywords History of ideas · Quantitative methods · Qualitative methods · Statistical analysis of textual data · Diachronic corpora · Scientific literature

1.1 Quantitative Methods and History of Ideas

Quantitative methods for the analysis of scientific literature have already been utilized by a variety of disciplines and the growing availability of large databases calls for fresh methods to deal with emerging problems, to open the door to different questions, and to lead to new knowledge.

The main aim of this book is to uncover the opportunities of learning the evolution of the ideas of a discipline through a distant reading (Moretti 2013) of the contents conveyed by relevant scientific literature. The temporal evolution of ideas (theories, concepts, methods, and application domains) has been explored by means of the temporal evolution of the occurrences of "words" (with particular reference to "keywords", e.g. technical words, scientific terms, proper names) included in papers published by mainstream, leading, scientific journals. Quantitative methods, statistical techniques, and software packages are used to identify and study the main subject matters, both in the past and today, of a discipline from raw textual data. From a theoretical viewpoint, the book also aims at dealing with a concept of "quality of life" of words over time and at fostering a debate about the popularity of ideas rather than dealing merely with the problem of dating their birth (that represents one of the main concerns in the study of the history of languages).

The experiment is innovative because quantitative methods for text analysis have been used in areas of human and social sciences which are traditionally explored through qualitative approaches. The chapters show that from the point of view of different areas (philosophy, sociology, psychology, linguistics, and statistics), it is possible to obtain an effective (distant) reading of large amounts of scientific articles and that quantitative methods can work successfully alongside qualitative methods in the study of the history of a discipline. However, we are aware that these achievements represent only a first step in an immense, boundless field, and much work remains to be done.

This book reports a new development of a research study conducted by a small group of Italian scholars who worked together on an interdisciplinary research project funded by the University of Padova, *Tracing the History of Words. A Portrait of a Discipline Through Analyses of Keyword Counts in Large Corpora of Scientific*

Literature, that considered the analysis with quantitative methods of corpora of scientific literature. Following a preliminary study on the history of statistics (Trevisani and Tuzzi 2015), the project dealt with philosophy and sociology. In a second stage, the research group grew up and embarked in a new project: included further disciplines (social psychology and Italian linguistics), exploited new methods for the analysis of textual data, and it is still going on in the effort of joining new groups.

Those who study the history of ideas of a discipline (e.g. the history of philosophical ideas, history of sociological ideas) usually rely on a long tradition of research that brings together the most influential and authoritative readings of what was in the past and what is now the life cycle of ideas. In all areas of knowledge, a history of ideas and a narrative of what has been the evolution of the disciplines have been developed from a theoretical, epistemological, and methodological point of view. But what normally happens is that this history becomes, over time, a history told in the light of what is known ex-post and, as a result, today we have a "representation" of the history of the milestones experienced by a discipline that has been reworked, revised, and corrected in the light of the results that the proposed ideas have had in subsequent years. Furthermore, prominent scholars of the past have in particular been the subject of studies on the history of disciplines, who certainly have a following, but when the focus is on the ideas of the "Great Names", a great deal of the brainwork that has prepared the ground for great discoveries is overlooked.

At least for the major contributions of contemporary scientific and intellectual panorama, there is an alternative option of reading the history of a discipline through the works published in journals. Scientific journals are the new Agora for the exchange of ideas and for the dissemination of research results, and because they represent a written and documented legacy, it is possible to read the timings of scientific debate across the temporal sequence of the publications.

In this book, we propose a reading of the history of some disciplines through a distant reading of the contents conveyed by the articles published in mainstream journals. We are aware, in turn, that this is not "the" representation of the history of a discipline, but a narration from a particular viewpoint, which reflects the historical moment in which the journals under consideration were published. It is well-known that not everything that has been published has the same importance in the history of a discipline and the same influence in the instruction of future generations, but what is published leaves a trace and to some extent has been taken into account by the scientific community of that specific historical moment. We also know that scientific discoveries and innovative ideas are not published when they are brought to light by their creators but only with some delay, that is, after they have been accepted and evaluated by the scientific community of that historical moment and after considerable publication times. In addition, authors did not begin their work when it becomes published (especially when it comes to mainstream journals) and even the most famous scholars may have struggled to find time at the beginning of their careers or when they had proposed ideas too innovative, original, or outside the box.

This way of reading the history of a discipline allows us to bring out what at a specific time was deemed relevant, either because it was a trend at that time, or because it was lauded by the editor and/or board of the journal, or because it was deemed of a high level by referees, etc. What we present is, therefore, an identification of the main ideas (theories, concepts, methods, and fields of application) that, at the time of publication, were relevant to the most influential journals and the dominant scientific communities linked to them.

1.2 Tracks on the Ground: What Methods for What Purposes

Grounded in a field that is closed to the perspective of distant reading (Moretti 2013), the projects exploited computational methods for text analysis and created a shared theoretical and practical framework to achieve innovative data-driven findings across different disciplines. Since this research group operated in an interdisciplinary framework, the state of the art cannot be "just one" as it is essentially specific of each discipline and of each approach. As a consequence, the traits to draw a general background of a desirable link among quantitative methods, qualitative methods, and history of ideas will be tackled through the chapters of this book with specific reference to each discipline. From a methodological viewpoint, we can try to trace only a sort of general and brief background of the methods adopted, without any pretence of completeness.

Even though quantitative linguistics enjoys a long tradition, the "modern" idea of quantitative analysis of textual data (ATD) emerged in the 1980s (Beaudouin 2016; Bolasco 2005, 2013; Lebart et al. 1998). A number of scientific and cultural approaches as well as theoretical schools and research instruments have developed since then and today a sheer size of research fields are hard to distinguish and systematize (e.g. see Wang et al. 2018; Léon and Loiseau 2016; Kelih et al. 2016; Tuzzi et al. 2015; Mayaffre et al. 2016; Mikros and Mačutek 2015; Née et al. 2014; Obradović et al. 2013; Naumann et al. 2012; Dister et al. 2012; Köhler 2011, 2012; Popescu et al. 2009; Popescu 2009; Baayen 2001). Moreover, branches of research that today are recognized as separate disciplines (quantitative linguistics, computational linguistics, text mining, stylometry, digital methods for text analysis, etc.) have some common roots and over time they differentiated in terms of methods, aims, and objects of research. During the last decades, research activities in this field have experienced a rapid development, and this process has fostered a renewed interest for topics related to text analysis and new methods to achieve a distant reading of large amounts of texts. Many approaches for text mining (Aggarwal and Zhai 2012; Berry and Kogan 2010; Berry 2004; Sanger and Feldman 2007; Kao and Poteet 2007; Sahami and Srivastava 2009; Sullivan 2001; Weiss et al. 2005) combine linguistic concepts, computational methods, information technologies, statistical learning, and machine learning to analyse texts. The field is highly interdisciplinary and it is constantly growing.

A diachronic corpus is a collection of texts including information on their timings (e.g. the publication date of an article). Scientific journals represent a useful ground for studying the development of scientific language and topics since we can assume that the evolution of word occurrences reflects the evolution of the corresponding concepts (Trevisani and Tuzzi 2015, 2018; Popescu and Strapparava 2014; Chavalarias and Cointet 2008, 2013; Guérin-Pace et al. 2012; Hall et al. 2008; Salem 1988, 1991). In quantitative linguistics, a number of textual features can be observed as sequences of linguistic properties (Mikros and Mačutek 2015; Köhler and Galle 1993) and the problem of reading the evolution of a phenomenon over time is often tackled by resorting to linguistic laws (Köhler 2011; Tuzzi and Köhler 2015) or time series analysis (Pawłowski 2006, 2016; Pawłowski et al. 2010). From a statistical viewpoint, a word trajectory hardly shows a regular behaviour and requires special attention since in diachronic corpora data are typically sparse over time (an unavoidable feature of textual data known as the "large p, small n" problem; see, for example, Hastie et al. 2008; Tibshirani et al. 2015; Johnstone and Titterington 2009; Lebart et al. 1984).

When diachronic corpora are collections of scientific literature reference can be made to methods based on scientometrics (see, for example, Yin and Wang 2017; Cobo et al. 2011, 2012; Porter and Rafols 2009; Small 2006) and also methods for content mapping based on occurrences and co-occurrences of words (see, for example, Guérin-Pace et al. 2012; Tuzzi 2012; Maggioni et al. 2009; Cretchley et al. 2010; Michel et al. 2011; Van Den Besselaar and Heimeriks 2006; Cahlík and Jiřina 2006; Bhattacharya and Basu 1998) and clustering for assessing significant changes (Zhang et al. 2016, 2017; Koplenig 2017; Gries and Hilpert 2008, 2012; Hilpert and Gries 2009; Diwersy and Luxardo 2016) prove useful. A relevant research area is topic modelling, that starting with the seminal work of Blei and Jordan (2003), has been further developed by Griffiths and Steyvers (2004) that introduced a Latent Dirichlet Allocation (LDA) generative model to discover topics covered in the corpus (Hall et al. 2008). Topic modelling connects to scientometrics and an interesting overview, also from an epistemological perspective, has been provided by Chavalarias and Cointet (2008, 2013). In this volume also an alternative way for identifying topics provided by Reinert's method (Ratinaud and Marchand 2012; Reinert 1983, 1990, 1993) and mainly developed in social sciences is exploited for the analysis of scientific literature.

In order to shape the history of individual words, a functional data analysis approach (Ramsay and Silverman 2005) is adopted and clustering methods for functional data are used to identify groups of keywords portraying similar temporal patterns. Two approaches to curve clustering are, in principle, viable: model-based and distance-based. The former is usually founded on finite mixture models and Gaussian processes for distributions (James and Sugar 2003; Jacques and Preda 2014a) although mixed effects models (Coffey et al. 2014; Giacofci et al. 2013; Trevisani and Tuzzi 2015) and non-Gaussian distributions (Lee and McLachlan 2013) or, within the Bayesian framework, Dirichlet processes (Angelini et al. 2012; Rodriguez et al. 2009; Ray and Mallick 2006) can be assumed for mixture components. In this volume, we opted for a distance-based approach as one of our objectives was to set up an exploratory and mostly automated learning procedure to be integrated in a so-called knowledge-based system (Trevisani and Tuzzi 2018), that is a computer system capable of generating knowledge by a large-scale integration of data, information as well as knowledge from different sources (linguistic and specific subject matter expertise), and endowed with a user-friendly interface. Within distance-based methods k-means type clustering algorithms have been widely applied to functional data especially when combined with finite basis-expansion approaches. Further choices, which extend the classical k-means algorithm with functional data, are also available (see Jacques and Preda 2014b; Wang et al. 2016).

1.3 Tracing the History of Words: A Quantitative Way

The quantitative perspective adopted by this research is essentially based on words and word counts (i.e. it is lexical based and refers to a "bag of words" approach), and, in particular, on the presence, absence, and occurrence over time of keywords relevant to the study of a specific discipline. Occurrence is an imperfect measure of the relevance of a word, however, with regard to scientific journals, we know to handle a textual genre in which language tends to be precise and succinct. In particular, titles and abstracts of the articles are extremely short, thick, and concise: They include keywords, scientific terms, technical words, nouns (e.g. research objects), proper names (e.g. authors), and often nothing else. Consequently, the fact that certain words are present or absent, and that they occur more frequently in certain historical periods and rarely in others gives us important information on the evolution of ideas and also on how to represent them.

All of text corpora exploited in the contributions of this book are written in English or Italian but the proposed methods can be extended for applications to any other language. However, each language envisages specific technical measurements and precautions that are heavily language-dependent, particularly to fulfil the phases of text pre-processing and processing (tokenization, cleaning, identification of multiword expressions, part of speech tagging, etc.).

An important assumption of this research is that the temporal course of word occurrences is viewed as a proxy of word diffusion and vitality, i.e. a word's life cycle. We assume, therefore, that the individual trajectories of words reflect the relevance through time of the corresponding ideas in the scientific discourse. Moreover, the research projects aimed at achieving interpretations of these findings. The fact of wanting to observe the trajectories drawn over time by occurrences of words also opens an interesting theoretical perspective that concerns the study of the difference between the first occurrence and the "settlement" of a given word. The research objective is not only to date the birth of subject matters but also to study their "fortunes" and fates. Moreover, to introduce the unprecedented concepts of "quality of life" of words and "life cycle" of ideas. The idea of "shaping the history of words" (Trevisani and Tuzzi 2015, p. 1288) is markedly unusual in linguistics and the study of the history of a language. Research in these areas focuses on the problem of dating the birth (first appearance) of a word and to study the possible semantic shift. Rarely do they care about the fate, or the eventual disappearance of a word.

Research has faced partially unexplored territory and has been shown to have great potential both from a theoretical point of view and from that of the application fields. Textual data retrieved from large corpora pose interesting challenges for any data analysis method and today represent a growing area of research in many fields. New problems emerge from the growing availability of large databases and new methods are needed to retrieve significant information from those large information sources.

1.4 Objectives and Procedures

As previously stated, the quantitative analysis adopted by these research studies is essentially based on words and word counts as part of the "bag of words" approach and, in particular, is based on the occurrences over time of the most significant keywords for the study of a specific discipline.

By *occurrence*, reference is being made to the number of repetitions of a word in a corpus of texts, usually expressed as the relative frequency (or rate) as compared to the size of the texts.

By *keywords*, reference is being made to a set of words (e.g. *theory*) and of word sequences (e.g. *theory of knowledge*) that have been identified by means of specific automatic (or semi-automatic) recognition procedures relevant to the study of a specific discipline. The keywords represent theories, concepts, scientific terms, technical terms, proper names, etc.

The work carried out for each of the disciplines involved in the projects pursued the following objectives:

1. To select relevant journals and compile suitable text corpora,
2. To obtain a first overview of the relationship between time and contents in order to verify the existence of a latent temporal pattern (correspondence analysis),
3. To identify relevant keywords,
4. To cluster keywords portraying similar temporal patterns and to identify latent dynamics of cluster keywords (curve clustering),
5. To identify relevant topics as groups of related words (topic detection).

1.4.1 Selection of Journals and Corpus Description

First of all, the experts of the disciplines selected journals to work on taking into account their reputation and centrality to the discipline. When possible, it was decided to go back to the year of the journal's founding. The texts were collected in a corpus (see Chap. 7) through a phase of text harvesting, which consists of downloading information (authors, title, year, volume, issue, number of pages and, if available, abstract) from public websites of these journals, through repositories and also resorting to printed versions. It was necessary to merge several sources because those available are not always complete and accurate as you would expect and it was necessary to look for other sources and resort to printed versions of journals to be able to fill the gaps and to collect all items. For some insights, it was decided to work with selected articles in full text (see Chaps. 2 and 5), also retrieved from the printed version of the journals.

These first raw data were processed to obtain a detailed overview of the available material, for information about the period of observation and to get a description in terms of the number, frequency, regularity and size of volumes, and issues of the journal. Similarly, the number, frequency, regularity and size of the articles, titles, and abstracts were examined.

In this phase, depending on the discipline being studied, decisions on the possible selection of items were also taken. It should be kept in mind that the journals not only publish scientific articles in the strict sense, some of the items retrieved from archives are not articles (e.g. *List of publications, News, Reviews*); some of them do not include content words in the title or in the abstract (e.g. *Comment, Rejoinder*) and, since many of them are works from the past, often they do not have abstracts.

At the end of the text harvesting, there is a diachronic corpus, i.e. a collection of texts including information on their time period, e.g. the publication date of an article. These texts might be arranged into groups (subcorpora) that refer to the same time interval, thus generating a temporal sequence of text sets. In our case, for each different journal we have diachronic corpora of texts (titles, abstracts, full texts) that are grouped by volume, which usually corresponds to groups by year of publication (with some exceptions in which volumes and years do not coincide).

A first assessment of the size of the corpora was made on the basis of words count "as they appear in the texts" (in technical terms this is called *forms* or graphical forms), that is, words are defined simply as sequences of characters of the alphabet isolated in the text by means of separators (spaces and punctuation). The recognition phase of words in the texts is technically called "tokenization" and is followed by a phase of "cleaning" that, in our specific case, essentially consists in removing the upper case and in the recognition of proper names. Other forms of tagging are used in the later phases, for example, the "part-of-speech" (POS), which serves to assign to each word, the lemma, and the grammatical category, or the "stemming", which serves to attribute more words to the same root (or "stem").

Once recognized and counted, the words can be divided into occurrences, or word-tokens, and in distinct words, or word-types. The frequency of a word-type is the number of corresponding word-tokens, the number N of word-tokens is the size of the corpus in terms of occurrences, the number V of word-types is the size of the corpus in terms of different words and the set of word-types represents the vocabulary (or word list) of the corpus. The observation of the word list and frequencies leads to first considerations of the most frequent words of the corpus. In addition, at the end of this phase it is possible to make some initial assessments on the length of available titles, abstracts, and articles.

1.4.2 Correspondence Analysis (CA)

Correspondence analysis (CA) is an explorative data analysis (EDA) that can be used to create content mapping and that, in this research, has served to reconstruct the general system of relationships among years, among words and between years and words (Greenacre 1984, 2007; Murtagh 2005, 2010, 2017; Lebart et al. 1984, 1998). The CA is based on the word list (vocabulary) that for each word comprises occurrences in the different volumes/years, representing our reference time-points. It recognizes similarities and differences through the lexical profiles, that is, through the frequencies of words over time: two words are similar (and close on the graph) because they have been used with similar frequency in the same time-points (volumes); two time-points are similar (and close on the graph) because the volumes of the journal at that time used the same words with similar frequency.

Although the CA is not able to definitively describe all relevant features of a large corpus, exploring relationships between words and years contributes to obtain an effective and immediate (distant) reading of the main contents and to distinguish features otherwise hardly perceptible with the sequential reading (close-reading) of texts. In the graph generated by the CA, it is possible to immediately verify whether the volumes of the journal have experienced an evolution of the contents over time. In fact, if the journal had a clear chronological development its volumes can be seen as a line on the plane that respects the order in time (see, for example, Fig. 2.1, Chap. 2). Only if a journal displays a chronological development of its contents does it make sense to use it for the study of the history of ideas of a discipline. For this reason, we considered this first EDA as consistent with our aims. For our processing, the words that exceeded a frequency threshold in the vocabulary of the corpora were selected. The threshold chosen was based on the coverage rate of the corpus and the coverage rate of the vocabulary.

Correspondence analysis is still widely used in the analysis of textual data. Since CA offers a way to achieve a Euclidean embedding of different information spaces based on cross-tabulation counts, it handles multivariate numerical and symbolic data with ease and proves useful to analyse great masses of textual data. From this perspective, it can be exploited in information semantics, and particularly in "big data" settings, to collect relationships. Murtagh (2010) showed how to work with CA as a Semantic Analysis Platform and through further experiments (Murtagh 2017) that involved data analysis in very high-dimensional spaces showed the benefits of CA as a tool suitable for carrying out latent semantic or principal axes mapping in big data scaling.

Since CA is a well-known and established method, this book does not include a specific chapter on this topic (see Greenacre 1984, 2007; Murtagh 2005; Lebart et al. 1984, 1998). However, the Appendix of this chapter provides a brief introduction to understand the rationale of CA.

1.4.3 Identification of Keywords

> "Within the perspective of analysis of textual data and, more in general, in all cases of data collection based on text harvesting, the construction phases, first of the corpora and then of textual data, are essential moments: choices made before statistical analysis are crucial to guarantee the quality of data" (Trevisani and Tuzzi 2015, p. 1289).

There are many examples of statistical analysis of textual data in literature that simply take into account the most frequent words in the corpus (or the most frequent n-grams, n-word-grams, etc.) and grammatical words are usually excluded from the word sets to be analysed (often referred to as "stop words"). We preferred not to follow these procedures systematically because they do not sufficiently take into account redundancies, compounds, and ambiguities. We preferred to retrieve all the relevant keywords in semi-automatic mode and identify content words and sequences of words (compounds, multiword expressions, segments) that are relevant to the study of a discipline. To this end, we also adopted a procedure for the automatic recognition of multiword expressions (MWEs, see Chap. 8) as well as the intersection of the corpus vocabulary with discipline-specific glossaries, and, when it was possible, also the index of keywords for the retrieved papers available in the online databases. It is worth mentioning that in any text we have sequences of words that have different meanings if they are considered alongside adjacent words, i.e. when we read them from a keyword in context (KWIC) perspective. The observation of the occurrences of MWEs increases the amount of information conveyed by keywords because a sequence of words partly reduces the noise and disambiguate the meaning. Moreover, semantic changes and semantic shifts of a word over time should envisage also the appearance of new MWEs and, when these new objects become relevant in a scientific language, their occurrences in publications should increase as well (and they should be retrieved by our procedures). We select nouns, names, MWEs, and through a matching with the most appropriate glossaries of each discipline we verified whether we collected all relevant keywords.

1.4.4 Curve Clustering

The purpose of curve clustering is to cluster keywords portraying similar temporal patterns and to identify latent dynamics of cluster keywords. This approach assumes that the "shape" of each keyword's trajectory in the volumes of mainstream journals, as it has been drawn by the keyword's occurrences over time, reflects the relevance of the corresponding subject matters in the scientific discourse.

In the frame of functional data analysis (FDA) approach, the proposed method consists of a two-stage functional clustering approach for statistical learning: first, a filtering step in which functional data are represented as smooth functions by a basis-expansion method (B-splines), second, a distance-based curve clustering in which the k-means algorithm is used combined with a metric to measure the distance between curves. Before filtering, a crucial choice concerns how to properly normalize word raw frequencies. Lastly, interpretation by expert opinion to decipher detected dynamics and thus compose a narrative of the evolution of the discipline is the conclusive step of the learning process (see Chaps. 6 and 9).

The main difference between this approach and the one represented by topic detection methods is that this analysis seeks to study the individual micro-histories and identify groups of keywords that, regardless of whether they belong to similar or completely different topics, have experienced the same temporal evolution over time. The main advantage is being able to effectively visualize trajectories that may represent interesting developments: words that, over time, have shown a growing trend, words that were in vogue in the past and have shown a decreasing trend, words which enjoyed a golden age in a period of great popularity and then disappeared, and words that live alternating phases of presence and absence, etc. (see, for example, Fig. 6.3, Chap. 6; Fig. 9.6, Chap. 9).

For our computations, all the keywords with a high enough frequency to produce a distinguishable trajectory and in a large number (but limited, usually to around 1000) were chosen to represent all the most relevant subject matters of a journal.

1.4.5 Topic Detection

Topic detection methods have objectives and assumptions very different from those of curve clustering. In this approach, the topics are identified as lists of related words that have in common the fact of co-occurring in texts. From a diachronic perspective, the temporal evolution of a topic does not depend on the trajectories of the words that compose it (i.e. taken individually) but by the "weight" that the frequencies of these words (as a whole) have comprehensively in different periods of time. The main advantage of topic detection with respect to curve clustering is that this approach makes it possible to automatically identify which are the main topics that emerge in the corpus without having to select a priori a list of keywords upon which to concentrate the analysis.

For our purposes, we used two different methods: Reinert's method and Latent Dirichlet Allocation (LDA). The two methods are both valid and, in many cases, have been preferred only on the basis of better readability of the results (see Chap. 10). Reinert's method is based on occurrences of words in texts and a similarity measure (chi-square distance). A descending hierarchical cluster analysis is performed on a distance table, which generates classes of units that best differentiate the vocabulary: it extracts classes of words that co-occur and that are best differentiated from other classes. Latent Dirichlet Allocation (LDA) depends on a topic-modelling algorithm, i.e. it bases itself on a generative statistical model that assumes the existence of a generative process: documents are generated by first picking a distribution over topics; words are generated by picking a topic from this distribution and then picking a word from that topic. For modelling the contributions of different topics to a document, LDA treats each topic as a probability distribution over words, viewing a document as a probabilistic mixture of these topics.

1.5 Chapters Outline

The volume is divided into two parts: there are five chapters in the first part for each of the disciplines involved (Chap. 2 philosophy, Chap. 3 sociology, Chap. 4 psychology, Chap. 5 linguistics, and Chap. 6 statistics) with the results of the analyses of corpora; the second part is dedicated to four methodological insights, which describe the methods used for processing the data (Chap. 7 compiling and pre-processing corpora, Chap. 8 MWE identification, Chap. 9 curve clustering, Chap. 10 topic detection). A concluding chapter summarizes the main findings and, perhaps first and foremost, the challenges of this interdisciplinary research work (see Chap. 11). Although the chapters of the first part of the book discuss the history of different disciplines and the contents of several journals, they are very similar to each other from the point of view of the approach and methods adopted because the work within the research group had been coordinated.

All contributions were primarily focused on verifying the existence of temporal patterns in the chronological textual data sets (diachronic corpora) and demonstrating by statistical analysis that the evolution of the contents of journals actually follows a chronological pattern. In fact, although it may seem a reasonable a priori condition, the existence of a temporal pattern in the evolution of the contents of a journal cannot be taken for granted and should be verified through processing of the data. In fact, a corpus which is diachronic in the structure does not always show a clear temporal pattern of its contents (Cortelazzo and Tuzzi 2007) and even a few exceptions were found in this research. For example, when we dealt with journals that arrange all their publications into special issues (that are focused on specific topics).

For the study of the temporal evolution of the contents of journals, methods to study the temporal trajectories of individual words were used as well as methods for topic detection. Depending on the discipline, the results obtained were compared with the ("content-metric") analysis of several journals. The possibility to work with titles, with abstracts, or with full-text articles was evaluated on a case-by-case basis with a focus on habits and scientific writing traditions of the relevant discipline and with reference to practices and traditions of the specific journal subject to analysis.

1.6 About This Book

In our research activities, we often have to take into account a large number of scientific papers in order to trace the history of a discipline and the temporal development of ideas in a specific field. There are definitely too many texts for one scholar to read in a lifetime. Instead of close-reading a limited number of texts, we have the opportunity to work with thousands of texts, uploading them into the memory of a computer and ask a software package to produce analyses and results. A software package (and a statistical model behind it) cannot "close read" a text. On the contrary, by means of mathematical and statistical tools, it might be smart enough to "distant read" a text, i.e. collecting data, retrieving relevant information, summarizing features, and finding patterns.

This book is a valid read to learn how quantitative methods can be part of the research instrumentation and the "toolbox" of scholars of humanities and social sciences. The decision to use a non-technical language makes the discussion accessible to those who do not have mathematical, computer, or statistical backgrounds. Nevertheless, some chapters are more technical as they deal with the specific requirements of statistical tools and discuss strengths and weaknesses of methods (see Chaps. 6 and 8). Since the analyses have been described with the views of the experts of the disciplines studied, those who have never used quantitative methods can see the results that can be achieved. The authors propose a convincing qualitative interpretation of the findings, results showed clear-cut temporal patterns in textual data, significant groups of words are proven to share the same temporal evolution, and results reveal timings and distinctive traits that either corroborate or contest the current representation of the history of disciplines. Moreover, the chapters offer an interesting look at what results are of the most interest to the disciplines involved and they show how the findings may be divulged and correlated with qualitative approaches.

Acknowledgements To the members of the research team and co-authors of this book, which I had the honour to lead and coordinate, go all my respect and gratitude for having chosen to follow me in this challenging adventure and to join the small group of brave researchers who for some time shared my interest in this matter. I would like to recognize the open minds of our most senior colleagues, and their vision and desire to get involved on truly exceptional, unfamiliar terrain, and I am very satisfied with the work of my younger colleagues for the desire to learn which they have shown, and for the great enthusiasm that they dedicated to the project and for having become the real "research engine" of the group.

Appendix

A Brief Overview on Correspondence Analysis

Correspondence Analysis (CA) is an Explorative Data Analysis (EDA) that has proven useful in studying the conjoint distribution of two (or more) categorical variables. CA portrays the existing structure of association between two (or more) variables by means of simple plots that position the categories of the variables on a plane.

The quantitative perspective adopted by the contributions of this volume are based on words and word counts, i.e. they are based on the observation of occurrences of relevant keywords over time. In this perspective, CA can be exploited to achieve a content mapping as it is useful to represent the system of relationships among years (e.g. volumes of the journals), among words (e.g. relevant keywords), and between years and words. Although CA is not able to describe all relevant linguistic features of a set of texts, it contributes to highlight latent patterns. For example, in our case, it makes it possible to verify whether the volumes of a journal expressed a clear temporal pattern in their main contents.

In the simplest version, CA works on a two-way contingency table in which the rows represent keywords (e.g. m word-types $w_1, \ldots, w_m$) and columns represent the volumes of the journal (e.g. p time-points $t_1, \ldots, t_p$). Each cell of this (lexical) contingency table represents the number n_{ij} of occurrences of the i-th keyword (the i-th row) in the volume published at the j-th time-point (the j-th column) (Table 1.1).

CA provides the best simultaneous representation of row profiles and column profiles on each axis (and on each plane generated by a pair of axes). The purpose of the CA is to translate the similarities between categories (words and volumes) in a graph in which the most similar categories are placed in adjacent positions in the space defined by the Cartesian axes. If you look at the words, it is fairly intuitive to think that the similarity between two words depends on how much the occurrences in the two rows of the table "resemble each other", that is, how similar they are in terms of presence, absence, or occurrence in the journal volumes: if two words tend to be used in the same volumes and with similar frequency, they have a similar profile over time. Two words with an identical profile will have no distance between them, that is, they will be represented on a graph as two overlapping points.

The intuitive notion of similarity between the profiles of two words w_i and w_k is translated into a distance (chi-square distance) that can be calculated for each pair of words:

$$d_{ik}^2 = \sum_{j=1}^{p} \frac{n}{n_{.j}} \left(\frac{n_{ij}}{n_{i.}} - \frac{n_{kj}}{n_{k.}} \right)^2$$

All the reasoning can be repeated by taking into consideration the similarity between pairs of volumes and considering the profiles of the two columns. Two volumes of the journal (time-points t_j and t_k) resemble each other if they have a similar lexical profile, i.e. if they include the same words with a similar relative frequency (Fig. 1.1).

The distance between two time-points t_j and t_k is given as:

$$d_{jk}^2 = \sum_{i=1}^{m} \frac{n}{n_{i.}} \left(\frac{n_{ij}}{n_{.j}} - \frac{n_{ik}}{n_{.k}} \right)^2$$

Table 1.1 Example of (lexical) contingency table words × time-points

	t_1	t_2	…	t_j	…	t_p	
w_1	n_{11}	n_{12}	…	n_{1j}	…	n_{1p}	$n_{1.}$
w_2	n_{21}	n_{22}	…	n_{2j}	…	n_{2p}	$n_{2.}$
⋮	⋮	⋮		⋮		⋮	
w_i	n_{i1}	n_{i2}	…	n_{ij}	…	n_{ip}	$n_{i.}$
⋮	⋮	⋮		⋮		⋮	
w_k	n_{k1}	n_{k2}	…	n_{kj}	…	n_{kp}	$n_{k.}$
⋮	⋮	⋮		⋮		⋮	
w_m	n_{m1}	n_{m2}	…	n_{mj}	…	n_{mp}	$n_{m.}$
	$n_{.1}$	$n_{.2}$	…	$n_{.j}$	…	$n_{.p}$	n

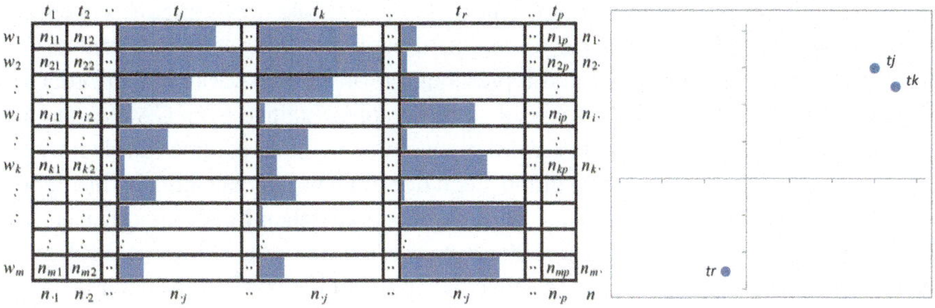

Fig. 1.1 Profiles in terms of relative frequencies and positions on the plane of three time-points

From another viewpoint, the rows and the columns of this matrix are considered as vectors, i.e. as points in a multidimensional space, and the distance between two vectors is measured through a weighted Euclidian distance that compares the corresponding lexical profiles taking into account the size of the subcorpora (volumes) at each time-point and the occurrences of each word in the corpus as a whole.

Following the calculation of the pairwise distance for words and for volumes, the next step is to transform the space generated by the original variables in a Euclidean space generated by new orthogonal variables (components or axes). The multidimensional space generated by the matrix is reduced to orthogonal dimensions (axes) that are displayed as Cartesian axes. The number of dimensions of this new space (i.e. the number of orthogonal axes) is equal to the number of linearly independent variables (rank of the matrix) that, in our context, is the number of time-points minus one ($p - 1$, more generally $min(m, p) - 1$).

The starting point of this transformation are the square matrix $m \times m$ which contains the pairwise distances between words and the square matrix $p \times p$ with the pairwise distances between volumes. The calculation of the coordinates of each axis is based on the singular value decomposition (SVD). The orthogonal factorial axes are sorted according to the amount of inertia collected (according to degree of association), i.e. they are in order of relevance: the first is the most important axis and the one which collects the highest portion of the information contained in the contingency table, the second axis is the one which collects the highest portion of information not explained by the first axis and so on. The Cartesian plane constructed with the first two factorial axes is the two-dimensional space which best represents the structure of association shown in the contingency table on a low-dimensional Euclidean space, and so on.

Unlike other analyses that move from the analysis of a matrix cases × variables, in CA the contingency table can be read in two ways: as m row vectors in the p-1 dimensions space generated by the columns, i.e. m words in the space of p time-points (volumes), and as p column vectors in the m-1 dimensions space generated by the rows, i.e. p time-points in the space of m words. From this observation, there is the immediate possibility to obtain two graphs separately: one with the words and one with the volumes. For the geometric properties of the two spaces (duality), the dimensions are the same and the two graphs overlap. This makes it possible to observe the system of relations between all the categories in play; although we must be very careful in the interpretation of the joint graphical representation of the two variables. In order to briefly summarize the elements for reading the graphs obtained from CA, we should remember that the position where a word or a volume is found assumes a role only in the globally created context of the graph, i.e. it doesn't have any meaning by itself, but it does have meaning in comparison with the positions taken by all the other points found in the solution with respect to the barycentre at the origin of the axes. If two words are close on the graph, it means that they have similar profiles and, analogously, if two volumes are close they have similar lexical profiles. The mutual position assumed by a word and a volume cannot be evaluated in a direct manner and must be evaluated with reference to the positions assumed by all the other elements. In this sense, it is useful to use the quadrants of the Cartesian plane and, thanks to the axes, the proximity can be evaluated by taking into account the angle formed by the axes (the more similar the angle formed with the axes is, the more they can be considered associated). The words or the volumes that contributed the most to the solution and which, therefore, can be considered the most important in the reconstructed context of the graph, are those which are distant from the origin of the axes. The densification of modalities in an area of the graph that stands out from the rest as a cluster might be interpreted as a semantic area and for this purpose one often choses to partition into clusters. The clusters of words or volumes should be homogeneous as much as possible within the group and, as much as possible, heterogeneous within groups. In the analysis of the lexical contingency table, a cluster analysis based on the CA groups together the volumes based on the lexical similarity (which is usually also visible in terms of proximity of the points on the graph).

An Example

To understand the functioning of the CA, an application example of a very simplified fictional corpus might be useful. Suppose you have 11 texts that include the topics of a journal of the statistical field and constitute a small text corpus:

text01 regression analysis; linear regression
text02 regression model; linear and non-linear model
text03 generalized linear model; parameter estimation
text04 sampling methods; random sampling; survey design and sampling methods
text05 survey design; finite populations
text06 methods for sampling elusive populations
text07 Normal distribution
text08 *z*-scores and Normal distribution
text09 Gamma distribution
text10 *p*-value: Normal distribution and Gamma—exponential family
text11 regression analysis; Normal distribution

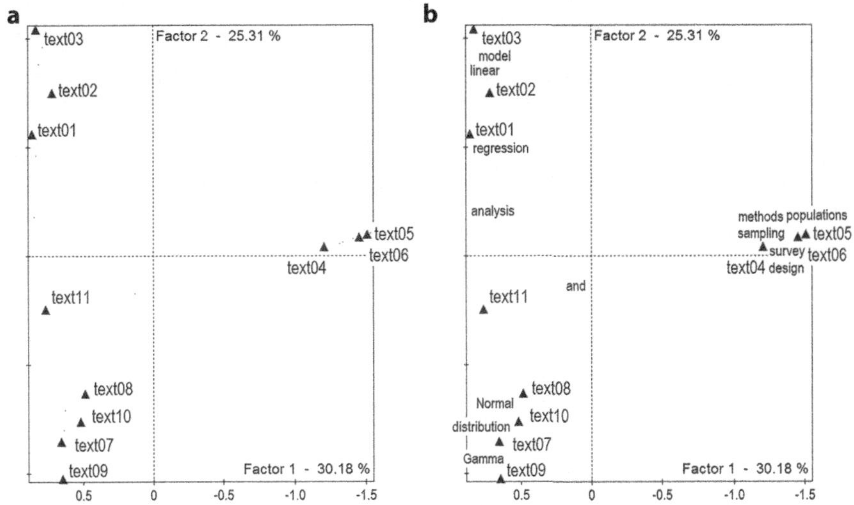

Fig. 1.2 First plane of correspondence analysis. Visualization of texts (**a**) and of both texts and words with frequency ≥2 (**b**)

There are 53 word-tokens and 25 word-types in the corpus. Taking into account only the words that are repeated at least twice, namely *distribution* (5 occurrences) *and, linear, Normal, regression,* and *sampling* (4), *methods* and *model* (3), *analysis, design, Gamma, populations,* and *survey* (2), we can construct a contingency table words × texts (Table 1.2), in which we see, for example, that the word *survey* was used once each by texts 04 and 05.

The CA of the contingency table results in 10 factorial axes. The first two axes collect 55% of the information (explained inertia) and the first factorial plane is shown in Fig. 1.2.

Figure 1.2 shows very well the three latent patterns present in the texts that refer to *linear model* (*regression, analysis*), *sampling methods* (*survey design, populations*), and *distribution* (*Normal, Gamma*). Texts 01, 02, and 03 can be found together in the area of *linear model* (second quadrant, upper left) while texts 07, 08, 09, and 10 in the area of *distribution* (third quadrant, bottom left). Text 11 is somewhere between linear models and distributions areas because it includes both topics. In the area of *sampling methods* (first quadrant, on the left), there are the texts 04, 05, and 06. It is interesting to note the conjunction *and* which is found near the origin of the axes because it has been used in different contexts (though slightly more often used by those who talked about distributions).

Table 1.2 Contingency table words × texts

Words	text01	text02	text03	text04	text05	text06	text07	text08	text09	text10	text11
distribution	0	0	0	0	0	0	1	1	1	1	1
and	0	1	0	1	0	0	0	1	0	1	0
linear	1	2	1	0	0	0	0	0	0	0	0
Normal	0	0	0	0	0	0	1	1	0	1	1
regression	2	1	0	0	0	0	0	0	0	0	1
sampling	0	0	0	3	0	1	0	0	0	0	0
methods	0	0	0	2	0	1	0	0	0	0	0
model	0	2	1	0	0	0	0	0	0	0	0
analysis	1	0	0	0	0	0	0	0	0	0	1
design	0	0	0	1	1	0	0	0	0	0	0
Gamma	0	0	0	0	0	0	0	0	1	1	0
populations	0	0	0	0	1	1	0	0	0	0	0
survey	0	0	0	1	1	0	0	0	0	0	0

Words with frequency ≥2

References

Aggarwal, C. C., & Zhai, C. (2012). *Mining text data*. New York: Springer.

Angelini, A., Canditiis, D. D., & Pensky, M. (2012). Clustering time-course microarray data using functional bayesian infinite mixture model. *Journal of Applied Statistics, 39*(1), 129–149.

Baayen, R. H. (2001). *Word frequency distributions*. Dordrecht: Kluwer Academic Publishers.

Beaudouin, V. (2016). Statistical analysis of textual data: Benzécri and the French School of Data Analysis. *Glottometrics, 33*, 56–72.

Berry, M. W. (Ed.). (2004). *Survey of text mining. Clustering, classification, and retrieval*. New York: Springer-Verlag.

Berry, M. W., & Kogan, J. (2010). *Text mining: Applications and theory*. Chichester: Wiley Online Library.

Bhattacharya, S., & Basu, P. K. (1998). Mapping a research area at the micro level using co-word analysis. *Scientometrics, 43*(3), 359–372.

Blei, D. M., Ng, A. Y., & Jordan, M. (2003). Latent Dirichlet allocation. *The Journal of Machine Learning Research, 3*, 993–1022.

Bolasco, S. (2005). Statistica testuale e text mining: alcuni paradigmi applicativi. *Quaderni di Statistica, 7*, 17–53.

Bolasco, S. (2013). *L'analisi automatica dei testi. Fare ricerca con il text mining*. Roma: Carocci.

Cahlík, T., & Jiřina, M. (2006). Law of cumulative advantages in the evolution of scientific fields. *Scientometrics, 66*(3), 441–449.

Chavalarias, D., & Cointet, J. P. (2008). Bottom-up scientific field detection for dynamical and hierarchical science mapping, methodology and case study. *Scientometrics, 75*(1), 37–50.

Chavalarias, D., & Cointet, J. P. (2013). Phylomemetic patterns in science evolution – The rise and fall of scientific fields. *PLoS One, 8*(2), e54847.

Cobo, M., López-Herrera, A., Herrera-Viedma, E., & Herrera, F. (2011). An approach for detecting, quantifying, and visualizing the evolution of a research field: A practical application to the fuzzy sets theory field. *Journal of Informetrics, 5*(1), 146–166.

Cobo, M., López-Herrera, A., Herrera-Viedma, E., & Herrera, F. (2012). SciMAT: A new science mapping analysis software tool. *Journal of the American Society for Information Science and Technology, 63*(8), 1609–1630.

Coffey, N., Hinde, J., & Holian, E. (2014). Clustering longitudinal profiles using P-splines and mixed effects models applied to time-course gene expression data. *Computational Statistics & Data Analysis, 71*, 14–29.

Cortelazzo, M. A., & Tuzzi, A. (Eds.). (2007). *Messaggi dal Colle. I discorsi di fine anno dei presidenti della Repubblica*. Venezia: Marsilio Editori.

Cretchley, J., Rooney, D., & Gallois, C. (2010). Mapping a 40-year history with leximancer: Themes and concepts in the journal of cross-cultural psychology. *Journal of Cross-Cultural Psychology, 41*(3), 318–328.

Dister, A., Longrée, D., & Purnelle, G. (Eds.). (2012). *JADT 2012 Actes des 11es Journées internationales d'analyse statistique des données textuelles*. Liège/Bruxelles: LASLA – SESLA.

Diwersy, S., & Luxardo, G. (2016). Mettre en évidence le temps lexical dans un corpus de grandes dimensions: l'exemple des débats du Parlement européen. In D. Mayaffre, C. Poudat, L. Vanni, V. Magri, & P. Follette (Eds.), *JADT 2016 - proceedings of 13th international conference on statistical analysis of textual data*. Nice: Pressess de Fac Imprimeur France.

Giacofci, M., Lambert-Lacroix, S., Marot, G., & Picard, F. (2013). Wavelet-based clustering for mixed-effects functional models in high dimension. *Biometrics, 69*(1), 31–40.

Greenacre, M. J. (1984). *Theory and application of correspondence analysis*. London: Academic Press.

Greenacre, M. J. (2007). *Correspondence analysis in practice*. London: Chapman & Hall.

Gries, S. T., & Hilpert, M. (2008). The identification of stages in diachronic data: Variability-based neighbour clustering. *Corpora, 3*(1), 59–81.

Gries, S. T., & Hilpert, M. (2012). Variability-based neighbor clustering: A bottom-up approach to periodization in historical linguistics. In T. Nevalainen & E. Traugott (Eds.), *The Oxford handbook of the history of English* (pp. 134–144). Oxford: Oxford University Press.

Griffiths, T., & Steyvers, M. (2004). Finding scientific topics. *Proceedings of the National Academy of Sciences of the United States of America (PNAS), 101*(Supplement 1), 5228–5235.

Guérin-Pace, F., Saint-Julien, T., & Lau-Bignon, A. W. (2012). The words of L'Espace géographique: A lexical analysis of the titles and keywords from 1972 to 2010. *Espace géographique, 41*(1), 4–31.

Hall, D., Jurafsky, D., & Manning, C. D. (2008). Studying the history of ideas using topic models. In *Proceedings of the Conference on Empirical Methods in Natural Language Processing*, 363–371.

Hastie, T., Tibshirani, R., & Friedman, J. (2008). *The elements of statistical learning: Data mining, inference and prediction* (2nd ed.). New York: Springer-Verlag.

Hilpert, M., & Gries, S. T. (2009). Assessing frequency changes in multi-stage diachronic corpora: Applications for historical corpus linguistics and the study of language acquisition. *Literary and Linguistic Computing, 24*(4), 385–401.

Jacques, J., & Preda, C. (2014a). Model-based clustering for multivariate functional data. *Computational Statistics & Data Analysis, 71*, 92–106.

Jacques, J., & Preda, C. (2014b). Functional data clustering: A survey. *Advances in Data Analysis and Classification, 8*(3), 231–255.

James, G. M., & Sugar, C. A. (2003). Clustering for sparsely sampled functional data. *Journal of the American Statistical Association, 98*, 397–408.

Johnstone, I. M., & Titterington, D. M. (2009). Statistical challenges of high-dimensional data. *Philosophical Transactions of the Royal Society A, 367*(1906), 4237–4253.

Kao, A., & Poteet, S. R. (Eds.). (2007). *Natural language processing and text mining*. London: Springer-Verlag.

Kelih, E., Knight, R., Mačutek, J., & Wilson, A. (Eds.). (2016). *Issues in quantitative linguistics 4. Studies in quantitative linguistics* (Vol. 23). Lüdenscheid: RAM-Verlag.

Köhler, R. (2011). Laws of languages. In P. C. Hogan (Ed.), *The Cambridge encyclopedia of the language science* (pp. 424–426). Cambridge: Cambridge University Press.

Köhler, R. (2012). *Quantitative syntax analysis*. Berlin: De Gruyter.

Köhler, R., & Galle, M. (1993). Dynamic aspects of text characteristics. In L. Hrebícek & G. Altmann (Eds.), *Quantitative text analysis* (pp. 46–53). Trier: Wissenschaftlicher.

Koplenig, A. (2017). A data-driven method to identify (correlated) changes in chronological corpora. *Journal of Quantitative Linguistics, 24*(4), 289–318.

Lebart, L., Morineau, A., & Warwick, K. M. (1984). *Multivariate descriptive statistical analysis: Correspondence analysis and related techniques for large matrices. Applied probability and statistics*. Chichester: Wiley.

Lebart, L., Salem, A., & Berry, L. (1998). *Exploring textual data*. Boston: Kluwer Academic Publication.

Lee, S. X., & McLachlan, G. J. (2013). Model-based clustering and classification with non-normal mixture distributions. *Statistical Methods & Applications, 22*(4), 427–454.

Léon, J., & Loiseau, S. (Eds.). (2016). *History of quantitative linguistics in France*. Lüdenscheid: RAM-Verlag.

Maggioni, M. A., Gambarotto, F., & Uberti, T. E. (2009). Mapping the evolution of 'Clusters': A meta-analysis. *FEEM* working paper no. 74.2009.

Mayaffre, D., Poudat, C., Vanni, L., Magri, V., & Follette, P. (Eds.). (2016). *JADT 2016 - Proceedings of 13th International Conference on Statistical Analysis of Textual Data, Nice 7-10 giugno 2016*. Nice: Pressess de Fac Imprimeur France.

Michel, J. B., Shen, Y. K., Aiden, A. P., Veres, A., Gray, M. K., The Google Books Team, et al. (2011). Quantitative analysis of culture using millions of digitized books. *Science, 331*(6014), 176–182.

Mikros, G. K., & Mačutek, J. (Eds.). (2015). *Sequences in language and text*. Berlin/Boston: Walter De Gruyter.

Moretti, F. (2013). *Distant reading*. London: Verso/New Left Books.

Murtagh, F. (2005). *Correspondence analysis and data coding with java and R*. London: Chapman & Hall/CRC.

Murtagh, F. (2010). The correspondence analysis platform for uncovering deep structure in data and information, sixth Boole lecture. *Computer Journal, 53*(3), 304–315.

Murtagh, F. (2017). Big data scaling through metric mapping: Exploiting the remarkable simplicity of very high dimensional spaces using correspondence analysis. In F. Palumbo, A. Montanari, & M. Vichi (Eds.), *Data science - innovative developments in data analysis and clustering* (pp. 295–306). Cham: Springer.

Naumann, S., Grzybek, P., Vulanović, R., & Altmann, G. (Eds.). (2012). *Synergetic linguistics. Text and language as dynamic systems*. Vienna: Praesens Verlag.

Née, É., Daube, J.-M., Valette, M., & Fleury, S. (Eds.). (2014). *Actes des 12e Journées internationales d'analyse statistique des données textuelles (JADT 2014), 3–6 juin 2014, Paris* (Actes électroniques).

Obradović, I., Kelih, E., & Köhler, R. (Eds.). (2013). *Methods and applications of quantitative linguistics: Selected papers of the VIIIth International Conference on Quantitative Linguistics (QUALICO)*, Belgrade, Serbia, April 16–19, 2012, Akademska Misao, Belgrado, Serbia.

Pawłowski, A. (2006). Chronological analysis of textual data from the Wrocław Corpus of Polish. *Poznań Studies in Contemporary Linguistics, 41*, 9–29.

Pawłowski, A. (2016). Chronological corpora: Challenges and opportunities of sequential analysis. The example of ChronoPress corpus of Polish. *Digital Humanities* (pp. 311–313).

Pawłowski, A., Krajewski, M., & Eder, M. (2010). Time series modelling in the analysis of homeric verse. *Eos, 97*(2), 79–100.

Popescu, I.-I., Macutek, J., & Altmann, G. (2009). *Aspects of word frequencies. Studies in quantitative linguistics*. Ludenscheid: RAM.

Popescu, I.-I. (2009). *Word frequency studies*. Berlin: Mouton De Gruyter.

Popescu, O., & Strapparava, C. (2014). Time corpora: Epochs, opinions and changes. *Knowledge-Based Systems, 69*, 3–13.

Porter, A. L., & Rafols, I. (2009). Is science becoming more interdisciplinary? Measuring and mapping six research fields over time. *Scientometrics, 81*(3), 719–745.

Ramsay, J., & Silverman, B. W. (2005). *Functional data analysis* (Springer series in statistics). New York: Springer.

Ratinaud, P., & Marchand, P. (2012). Application de la méthode ALCESTE à de “gros” corpus et stabilité des “mondes lexicaux”: analyse du “CableGate” avec IRaMuTeQ. In *Actes des 11eme Journées internationales d'Analyse statistique des Données Textuelles* (pp. 835–844). Liège, Belgique.

Ray, S., & Mallick, B. (2006). Functional clustering by bayesian wavelet methods. *Journal of the Royal Statistical Society: Series B (Statistical Methodology), 68*(2), 305–332.

Reinert, M. (1983). Une methode de classification descendante hierarchique: application a l'analyse lexicale par context. *Les Cahiers de l'Analyse des Données, 8*(2), 187–198.

Reinert, M. (1990). ALCESTE: Une méthodologie d'analyse des données textuelles et une application: Aurélia de Gérard de Nerval. *Bulletin de Méthodologie Sociologique, 26*, 24–54.

Reinert, M. (1993). Les “mondes lexicaux” et leur “logique” à travers l'analyse statistique d'un corpus de récits de cauchemars. *Language et Société, 66*, 5–39.

Rodriguez, A., Dunson, D. B., & Gelfand, A. E. (2009). Bayesian nonparametric functional data analysis through density estimation. *Biometrika, 96*(1), 149–162.

Sahami, A., & Srivastava, M. (Eds.). (2009). *Text mining: Theory and applications*. London: Taylor and Francis.

Salem, A. (1988). Approches du temps lexical. Statistique textuelle et séries chronologiques. *Mots. Les langages du politique, 17*, 105–114.

Salem, A. (1991). Les séries textuelles chronologiques. *Histoire & Mesure, VI-1*(2), 149–175.

Sanger, J., & Feldman, R. (2007). *The text mining handbook: Advanced approaches in analyzing unstructured data*. Cambridge: Cambridge University Press.

Small, H. (2006). Tracking and predicting growth areas in science. *Scientometrics, 68*(3), 595–610.
Sullivan, D. (2001). *Document warehousing and text mining: Techniques for improving business operations*. Wiley: Marketing and Sales.
Tibshirani, R., Wainwright, M., & Hastie, T. (2015). *Statistical learning with sparsity: The lasso and generalizations*. New York: Chapman and Hall/CRC.
Trevisani, M., & Tuzzi, A. (2015). A portrait of JASA: The history of statistics through analysis of keyword counts in an early scientific journal. *Quality and Quantity, 49*, 1287–1304.
Trevisani, M., & Tuzzi, A. (2018). Learning the evolution of disciplines from scientific literature. A functional clustering approach to normalized keyword count trajectories. *Knowledge-Based Systems, 146*, 129–141.
Tuzzi, A. (2012). Reinhard Köhler's scientific production: Words, numbers and pictures. In S. Naumann, P. Grzybek, R. Vulanović, & G. Altmann (Eds.), *Synergetic linguistics. Text and language as dynamic systems* (pp. 223–242). Vienna: Praesens Verlag.
Tuzzi, A., Benesová, M., & Macutek, J. (Eds.). (2015). *Recent contributions to quantitative linguistics*. Berlin: De Gruyter.
Tuzzi, A., & Köhler, R. (2015). Tracing the history of words. In A. Tuzzi, M. Benesová, & J. Macutek (Eds.), *Recent contributions to quantitative linguistics* (pp. 203–214). Berlin: DeGruyter.
Van Den Besselaar, P., & Heimeriks, G. (2006). Mapping research topics using word-reference co-occurrences: A method and an exploratory case study. *Scientometrics, 68*(3), 377–393.
Wang, J. L., Chiou, J. M., & Mueller, H. G. (2016). Functional data analysis. *Annual Review of Statistics and Its Application, 3*(1), 257–295.
Wang, L., Köhler, R., & Tuzzi, A. (Eds.). (2018). *Structure, Function and Process in Texts*. Lüdenscheid: RAM-Verlag.
Weiss, S. M., Indurkhya, N., Zhang, T., & Damerau, F. (2005). *Text mining: Predictive methods for analyzing unstructured information*. New York: Springer.
Yin, Y., & Wang, D. (2017). The time dimension of science: Connecting the past to the future. *Journal of Informetrics, 11*, 608–621.
Zhang, Y., Chen, H., Lu, J., & Zhang, G. (2017). Detecting and predicting the topic change of knowledge-based systems: A topic-based bibliometric analysis from 1991 to 2016. *Knowledge Based System, 133*(Supplement C), 255–268.
Zhang, Y., Zhang, G., Chen, H., Porter, A. L., Zhu, D., & Lu, J. (2016). Topic analysis and forecasting for science, technology and innovation: Methodology with a case study focusing on big data research. *Technological Forecasting and Social Change, 105*, 179–191.

Part I
Tracing the Life-Cycle of Ideas

Chapter 2
Tracing the Words of the Analytic Turn in the Journal of Philosophy

Giuseppe Spolaore and Pierdaniele Giaretta

Contents

2.1 Introduction 26
2.2 The Corpora and the Journal of Philosophy 27
2.3 Results 32
2.3.1 Title Corpus 33
2.3.2 Full-Text Corpus 36
2.4 Some Conclusions About the Analytic Turn 39
2.5 More General Issues 40
References 43

Abstract We carried out a quantitative analysis of the twentieth-century philosophical lexicon, based on corpora drawn from a leading US philosophical review, the Journal of Philosophy. In the chapter, we present and discuss some findings of ours, with special regard to analytic philosophy, a contemporary philosophical tradition that plays a leading role in the English-speaking world. More specifically, we present some results that are relevant to the process of rise to dominance of analytic philosophy in the USA—a process that we call "analytic turn". In a nutshell, we found that the lexicon related to topics that are characteristic of (half-of-the-century) analytic philosophy, such as semantics, logic and epistemology, underwent a significant gain in importance within the Journal of Philosophy during the three post-Second War decades. We also observed a corresponding decrease in importance of the lexicon related to non-English speaking, "continental" philosophy, and to themes of generic human and social interest. We also found that the turning point in both processes can be approximately located between the mid-1950s and the early 1960s. Although we shall discuss reasons to be cautious in drawing conclusions

G. Spolaore (✉) · P. Giaretta
University of Padova, Padova, Italy
e-mail: giuseppe.spolaore@unipd.it

A. Tuzzi (ed.), *Tracing the Life Cycle of Ideas in the Humanities and Social Sciences*, Quantitative Methods in the Humanities and Social Sciences,
https://doi.org/10.1007/978-3-319-97064-6_2

about the history of philosophy from lexical data concerning philosophical journals, these and other results of ours suggest a picture of the analytic turn that is at odds with some commonplace opinion among analytic philosophers.

Keywords Analytic philosophy · Analytic turn · Distant reading · Quantitative lexical analysis · Journal of Philosophy

2.1 Introduction

We studied the evolution of the philosophical lexicon during the twentieth century by means of statistical techniques, based on corpora drawn from a leading US philosophical review. Usually, the quantitative study of academic journals is bibliometric or "scientometric" in nature, that is, it focuses on (cross)references and other bibliographic data. In contrast, our study is "content-metric", that is, it hinges on the statistical analysis of the lexicon. To the best of our knowledge, it is the most extensive content-metric study ever carried out about contemporary academic philosophy.

In this chapter, we present and discuss some findings of ours, with special regard to the philosophical tradition called *analytic philosophy*. More specifically, we shall investigate the process of rise to dominance of analytic philosophy in US philosophy—a process that we arbitrarily call *analytic turn*. Moreover, we draw some historical and methodological considerations on the quantitative analysis of philosophical texts and its connections with more traditional, qualitative techniques in the history of philosophy.

The main corpus ("full-text corpus") we examined is formed of all the research papers published in the main section of the *Journal of Philosophy* (US, est. 1904) during the three post-Second War decades (1946–1975). We also considered a subsidiary corpus ("title corpus"), which includes the titles of all the contributions (including reviews) published in the *Journal of Philosophy* up to 2016.

Our main findings can be summarized as follows:

- There is evidence that, in the full-text corpus, the topics characteristic of (middle) analytic philosophy, such as semantics, logic and epistemology, undergo a significant gain in importance in the interval between the mid-1950s and the mid-1970s.
- During the same interval, it appears that the topics characteristic of "continental" philosophy, along with themes of generic human and social interest, are less and less represented as time goes by.
- The title corpus displays a recognizable chronological pattern in terms of lexical trends: volumes/years that are chronological close each other also mirror a similar lexical profile, i.e. they are similar as of their subject matters.
- A five-cluster analysis of the title corpus structures it into two broad groups (spamming, respectively, from the turn of the century to the 1960s, and from the mid-1960s to 2016) separated by a consistent hiatus.
- The clusters belonging to the second group (mid-1960s to 2016) are the only ones whose characteristic lexicon significantly reflects analytic trends and topics.

- In the title corpus, non-English words occur much more often in the earlier clusters than in the later ones.

For reasons that we shall partially discuss in this chapter, caution is needed in generalizing from linguistic data about a philosophical review to conclusions about the history of philosophy. However, there is a broad historical picture that naturally suggests itself on the basis of our findings, which receives independent support from other historical sources. Interestingly, this picture is at odds with some common notions about the rise of analytic philosophy.

During the 1940s, so the picture goes, (what would be later called) analytic philosophy did not occupy an especially preeminent position within the US philosophical scene. Analytic philosophy started to play a major role relatively late, well into the 1950s. Such analytic turn (viz., the process of rise to dominance of analytic philosophy) accompanied itself with two other processes: a process of progressive detachment from "continental" (or otherwise non-Anglo-Saxon) philosophical strands, and a process of specialization, with less and less attention paid to themes of general human and social interest.[1]

Let us look ahead. In the next section, we offer information about the corpora, and we briefly introduce our motivations in choosing them. In Sect. 2.3, we describe our main results and, in Sect. 2.4, we draw a few conclusions. Finally, in Sect. 2.5, we discuss some methodological issues related to the content-metric approach and its relations with more traditional historiographic methods.

2.2 The Corpora and the Journal of Philosophy

As mentioned above, our study concerns two corpora, the full-text corpus and the title corpus. Both corpora are drawn from the *Journal of Philosophy* (the Journal). In this section, we present these corpora and offer some historical information about the Journal. Contextually, we also provide justification for our choice of corpora.

The title corpus (see Table 2.1) is formed of the titles of all contributions published in the Journal up to (and including) 2016. The category variables available to the analysis are the volume number, the issue number, and the publication year. Each volume corresponds to a solar year. Overall, 113 Volumes have been published up to 2016. The number of issues for each volume changes through the years. From 1904 to 1963, each volume includes 26 issues; from 1964 to 1971, 24 issues; in 1972, 23 issues; from 1973 to 1976, 22. From 1977 onward, the Journal adopts a monthly publication rate, with 12 issues per volume. In some years, two issues are included into one and, in a single occasion, the same issue is divided in two. The

[1] However, it appears that during the twentieth century philosophical journals underwent a less marked process of lexicon specialization than journals belonging to other fields of inquiry, such as sociology, statistics, and psychology (see Chap.11).

Table 2.1 Journal of Philosophy, corpus of titles: volumes, issues containing at least one (relevant) title, number of titles and size in word-tokens (*N*)

Name of the journal	Years	Volumes	Issues	No. of titles	*N*
JPPSM*	1904–1913	1–10	260	1398	8678
JPPSM	1914–1923	11–20	259	1107	6644
Journal of Philosophy	1924–1930	21–27	182	786	4863
Journal of Philosophy	1931–1940	28–37	252	1768	11,623
Journal of Philosophy	1941–1950	38–47	260	1465	9120
Journal of Philosophy	1951–1960	48–57	258	1284	7955
Journal of Philosophy	1961–1970	58–67	246	912	5011
Journal of Philosophy	1971–1980	68–77	183	704	3682
Journal of Philosophy	1981–1990	78–87	121	640	3912
Journal of Philosophy	1991–2000	88–97	120	434	2894
Journal of Philosophy	2001–2010	98–107	120	352	2122
Journal of Philosophy	2011–2016	108–113	66	200	1281

Table 2.2 Lexicometric measures of the Journal of Philosophy (titles)

N—word-tokens	67,785
V—word-types	9917
$(V/N)*100$—type/token ratio	14.6
$(V_1/V)*100$—hapax percentage	59.3

total number of issues published up to 2016 and considered in the analysis is 2327. However, only the issues containing at least one title relevant to the analysis are included in the corpus.

The titles under analysis are 11,050, and they include both titles of articles and of book reviews. The corpus is formed of 9917 different word-types, with 67,785 occurrences. The number of hapaxes (i.e. types occurring only once in the corpus) is 5886, corresponding to 59% of the entries in the corpus vocabulary (Table 2.2).

An extract of the corpus vocabulary is shown in Table 2.3. "Empty" forms such as conjunctions or prepositions are ignored in the table, except for the two most common words (*the* and *of*), which are included for comparison. A rule before a form signals that some other forms have not been included in the table. Interestingly, the most common lexical items are *philosophy*, *study*, *theory*, and *logic*.

The full-text corpus is formed of all "featured" articles published in the Journal during the period 1946–1975. By "featured articles" we mean original papers published in the main section of the Journal, as opposed to summaries, discussion notes, reviews, news, advertisings, forewords, etc. Overall, the corpus embraces 1570 articles, distributed in 30 volumes and 748 issues, from issue 1 of volume 43 (1946) to issue 22 of volume 72 (1975). The average number of featured articles per issue is 2 (Table 2.4).

The total number of word-types is 97,897, with 7,314,328 occurrences and 50,659 hapaxes, corresponding to 51.74% of word-types (Table 2.5).

Table 2.3 Excerpt of contingency table words × years of the Journal of Philosophy (title corpus)

Words	Occurrences (corpus)	1904	1905	1906	:	1990	1991	:	2014	2015	2016
the	4,977	76	73	90	:	43	32	:	15	17	6
of	4,738	82	77	77	:	38	38	:	10	12	5
philosophy	654	3	6	6	:	5	5	:	1	1	0
study	359	4	9	6	:	3	2	:	0	0	0
theory	340	5	5	1	:	3	2	:	1	0	0
logic	272	6	2	2	:	2	5	:	0	3	1
science	254	5	3	5	:	1	1	:	0	0	0
psychology	240	8	9	8	:	0	0	:	0	0	0
ethics	218	6	2	1	:	1	3	:	0	0	0
value	214	1	0	0	:	0	1	:	1	0	0
moral	210	2	0	1	:	2	4	:	1	0	1
essay	205	0	3	3	:	1	0	:	0	2	1
mind	182	3	1	0	:	0	2	:	0	1	1
history	181	2	2	1	:	2	1	:	0	0	0
nature	177	1	1	4	:	1	0	:	0	1	0
knowledge	165	0	4	7	:	2	1	:	0	1	0
theory of knowledge	30	0	0	1	:	0	0	:	0	0	0
essence	30	0	1	0	:	0	0	:	0	0	0
philosophy of science	29	1	0	0	:	5	0	:	0	0	0
quality	29	0	1	0	:	0	0	:	0	0	0
significance	29	1	0	0	:	0	0	:	0	0	0
intelligence	28	0	0	0	:	0	0	:	0	0	0
America	28	0	0	0	:	0	0	:	0	0	0
pragmatic	28	1	0	0	:	0	0	:	0	0	0
discussion	28	0	0	3	:	0	0	:	0	0	0
cognitive	27	0	2	1	:	0	0	:	0	0	0
responsibility	27	1	0	0	:	1	1	:	0	1	0
Wittgenstein	26	0	0	0	:	1	0	:	0	1	0
moral philosophy	26	0	0	0	:	1	0	:	0	0	0
metaepistemology	1	0	0	0	:	0	0	:	0	0	1

Table 2.4 Journal of Philosophy, corpus full-texts: volumes, issues containing at least one (relevant) article, number of titles and size in word-tokens (*N*)

Years	Volumes	Issues	No. of articles	*N*
1946–1950	43–47	130	219	1,020,833
1951–1955	48–52	130	267	1,117,271
1956–1960	53–57	130	335	1,504,145
1961–1965	58–62	126	258	1,160,632
1966–1970	63–67	120	251	1,276,112
1971–1975	68–72	112	240	1,235,335

Table 2.5 Lexicometric measures of the Journal of Philosophy (full-texts)

(*V*) Word-types	97,897
(*N*) Word-tokens	7,314,328
$(V/N)*100$ = type/token ratio	1.3
$(V_1/V)*100$ = percentage of hapax	51.7

From a lexical viewpoint, it is interesting to have a glimpse of the most common forms (i.e. words, keywords, or multiwords; see Chap. 8) within the corpus, which are illustrated in Table 2.6. In Table 2.6, "empty" forms such as conjunctions or prepositions are ignored, except for the two most common words ("the" and "of") which are included for comparison. A rule before a form signals that some other forms have not been included in the table. The last rule marks the omission of all the forms preceding *superpropositional*, which has just one occurrence in our corpus. The most common lexical forms (such as *true*, *fact*, *theory*, *knowledge*, *experience*, *meaning*, *language*, *world*, and *argument*) are fairly natural words to be found within philosophical texts. It is worth observing, however, that none of them is a purely technical term from philosophy. As usual, multiword expressions have a more technical flavour, most noticeably *state of affairs*, *ordinary language*, and *philosophy of science*.

We started with articles as pdf files and converted them into txt files. As it happens with scanned documents, the automatic conversion produced less-than-ideal results, which had to be checked and corrected manually. This activity proved time consuming, and a 30-year corpus was the largest we could deal with given the resources at our disposal and the temporal limits. As for the 1946–1975 period, it struck us as a very natural choice, as we were primarily interested in the process of rising to dominance of analytic philosophy in the USA, and there are independent reasons to think that this process took place at some point within that interval.

Both corpora have been pre-processed by using TaLTaC2 software. They were normalized by replacing uppercase with lowercase letters. Multiword expressions were identified, manually checked and those with frequencies higher than or equal to 5 have been regarded as textual units. In order to recognize multiword expressions, an automatic information retrieval procedure was used (see Chap. 8). Further multiword expressions were found by means of the Blackwell Dictionary of Western Philosophy.[2]

The Journal was founded in 1904 as *The Journal of Philosophy, Psychology, and Scientific Methods* by Frederick J. E. Woodbridge (a philosopher) and J. McKeen Cattell (a psychologist). In 1906, Wendell T. Bush joined them as co-editor. It is worth emphasizing that, in its earliest phase, the Journal covered also psychological and other broadly scientific topics. Even now, it favours multidisciplinary topics and approaches, in accordance with its declared mission, i.e. to "encourage the interchange of ideas, especially the exploration of the borderline between philosophy

[2] http://www.blackwellreference.com/public/tocnode?id=g9781405106795_chunk_g97814051067952#citation. Accessed March 13, 2018.

Table 2.6 Excerpt of the contingency table words × years of the Journal of Philosophy (full-text corpus)

Words	Occurrences (corpus)	1946	1947	1948	:	1959	1960	:	1973	1974	1975
the	431,103	13,280	13,069	13,253	:	18,028	15,567	:	13,597	13,688	13,471
of	311,067	9461	9463	10,055	:	13,131	11,306	:	9427	9714	9354
true	10,565	153	281	175	:	322	352	:	465	451	454
fact	8650	264	274	249	:	334	340	:	274	231	276
theory	7093	160	168	135	:	226	128	:	329	268	387
knowledge	5619	239	240	278	:	324	108	:	210	188	54
experience	5559	190	409	359	:	237	160	:	39	52	45
meaning	5518	171	210	165	:	249	218	:	111	61	119
language	5253	109	184	80	:	323	164	:	145	100	160
world	5215	210	220	171	:	185	251	:	197	167	182
argument	5213	66	68	111	:	175	176	:	193	304	229
nature	5050	239	292	198	:	324	149	:	57	96	61
reason	4965	94	121	139	:	194	143	:	205	158	197
truth	4788	133	144	140	:	170	165	:	141	197	187
philosophy	4510	163	177	170	:	270	320	:	24	51	40
analysis	4248	107	107	124	:	159	146	:	145	104	136
relation	4232	157	138	126	:	248	82	:	247	106	149
problem	4202	119	163	128	:	175	179	:	133	116	148
existence	4052	147	162	117	:	117	254	:	113	83	80
mind	4046	186	139	200	:	165	119	:	95	129	51
order	3997	104	111	129	:	162	116	:	102	124	85
state of affairs	554	7	8	30	:	11	6	:	73	17	15
human beings	553	22	13	18	:	21	23	:	4	40	15
common sense	517	24	37	46	:	26	38	:	9	7	8
ordinary language	480	0	5	28	:	40	42	:	7	3	4
philosophy of science	436	12	15	3	:	22	8	:	21	19	16
superpropositional	1	0	0	0	:	0	0	:	0	0	0

and other disciplines".[3] The title of the Journal was shortened in its present form in 1923.

Historically, the editors[4] of the Journal are members of the Department of Philosophy at Columbia University—the Journal's publishing institution. In 1946, at the beginning of our main period of interest, the editors' panel included Herbert W. Schneider (a philosopher of religion, editor from 1923 to 1961), John H. Randall, Jr. (a historian of philosophy and ethicist, editor from 1937 to 1981), and Ernest Nagel (an extremely influential philosopher of science, editor from 1939 to 1956).[5] After Nagel left, two historian of (continental) philosophy started serving as editors, namely, Robert D. Cumming (from 1957 to 1964) and George L. Kline (from 1959 to 1964). It is worth observing that, with the possible exception of Nagel, none of these editors would have called himself an analytic philosopher. A major change in the editors' panel occurred during the mid-1960s, with the arrival of Arthur Danto (extremely influential art criticist and aesthetician, editor from 1964 to 2010), Sidney Morgenbesser (general philosopher and historian of philosophy, editor from 1964 to 1988), James J. Walsh (general philosopher and historian of philosophy, editor from 1964 to 1990), and Charles D. Parsons (philosopher of logic, mathematics and language, editor from 1966 to 1990). The last remarkable addition to the editors' panel before 1975 was Bernard Berofsky (ethicist and metaphysician, editor since 1970, still in charge). The distribution of editors during the 1946–1975 period is summarized in Table 2.7.

The Journal is one of the most esteemed peer-review philosophical journals of the world, being constantly rated among the top-five ones within the analytic tradition (see, e.g., Leiter 2015). Moreover, it is very ancient: other strong contenders such as *Philosophy and Phenomenological Research* (est. 1940) and *Noûs* (est. 1967) are considerably more recent and would not provide the same chronological depth to our study. Finally, we regarded the Journal as a slightly better choice than *Philosophical Review* (another very ancient and esteemed US philosophical journal) because it publishes more featured articles per year. However, we plan to extend our study to other journals in the future.

2.3 Results

In this section, we detail a number of results of our research that are related to the rise of analytic philosophy. Each result is accompanied with some words of comment, which concern either its interpretation or its connections with other results.

---

[3]As found in the site of the Journal, http://www.journalofphilosophy.org/generalinfo.html. Accessed March 13, 2018.

[4]When we speak of "editor", we always mean *full* editor, as opposed, e.g. to *consulting* editor, or *managing* editor, or *book-review* editor, and so on.

[5]We obtained information about the editors' panel directly from the list of editors published in the *Front matter* section of each issue of the Journal.

Table 2.7 Editors of the Journal of Philosophy, 1946–1975

Years	Editors' panel
1946–1956	H. W. Schneider, J. H. Randall, Jr., E. Nagel
1957–1958	H. W. Schneider, J. H. Randall, R. D. Cumming
1959–1961	H. W. Schneider, J. H. Randall, R. D. Cumming, G. L. Kline
1961–1964	J. H. Randall, R. D. Cumming, G. L. Kline
1964–1965	J. H. Randall, A. Danto, S. Morgenbesser, J. J. Walsh
1966–1969	J. H. Randall, A. Danto, S. Morgenbesser, J. J. Walsh, C. D. Parsons
1970–1975	J. H. Randall, A. Danto, S. Morgenbesser, J. J. Walsh, C. D. Parsons, B. Berofsky

2.3.1 Title Corpus

Correspondence analysis (CA, Greenacre 2007; Murtagh 2005; Lebart et al. 1984) is an established and well-known statistical method that provides a representation of volumes/years and words in a Cartesian plane. CA is a particular instance of the *principal component analysis* (PCA) of the rows and columns of a contingency table. It is thus applicable to a (words × years) table that includes occurrences.

The main aim of CA is that of turning the frequencies of words into coordinates on a multidimensional system of Cartesian axes: groups and words are represented in a low-dimensional space, by mapping a certain distance into a specific Euclidean distance (Murtagh 2005), which, in turn, is mapped into Cartesian planes. In our case, CA focuses on the lexical profiles of the years, and it is useful in highlighting the relations among years, among words, and among years and words. The relative positions of the years on the plane depend on their lexical similarity. From the viewpoint of an exploratory data analysis (EDA), CA proves very useful to highlight reciprocal positions of volumes/years and words and to verify the existence of a temporal pattern. The Cartesian system is also useful to exploit the Euclidean distance to find clusters of volumes/years endowed with similar lexical profiles (see Chap. 1).

In this analysis, we used a matrix of 1608 words over 113 years (rows per columns). Only words with frequencies higher than or equal to 5 were considered for the analysis. Keywords and multiword expressions have been labelled inside a column in the matrix set by a supplementary variable. In this way, it has been possible to set them apart from empty words, which has been omitted in the output. As mentioned above, based on CA, we found out that the title corpus displays a recognizable chronological pattern in terms of lexical profiles (see Fig. 2.1).

Moreover, the cluster analysis highlights the existence of a clear thematic separation starting in the early 1960s (Fig. 2.2).

Each cluster has an associated characteristic vocabulary, i.e. intuitively, the set of words that are more likely to be found within the cluster as opposed to the other clusters. These distinctive lists of words are listed in chronological order in Table 2.8.

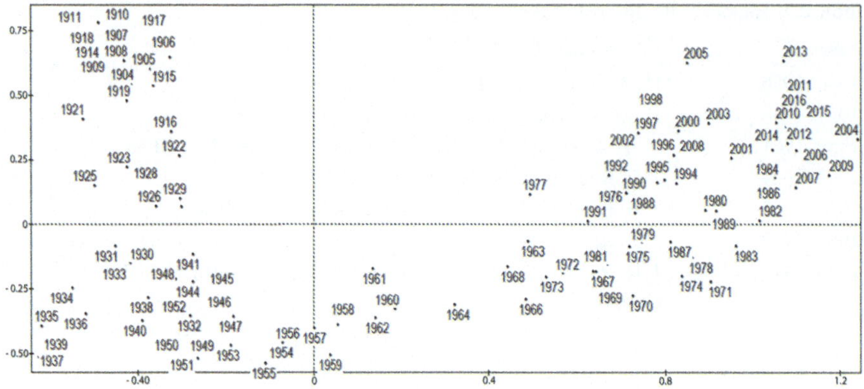

Fig. 2.1 First factorial plane of the Journal Correspondence Analysis. Projection of years

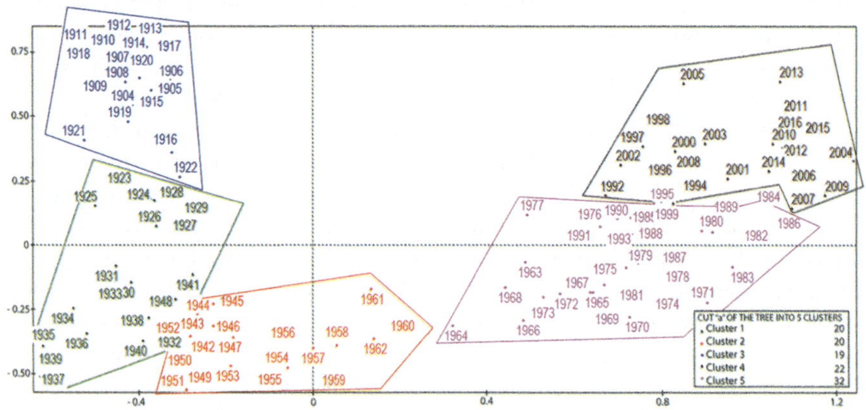

Fig. 2.2 First factorial plane of the Journal Correspondence Analysis. Projection of years divided into five clusters

The vocabularies are clearly differentiated from one another, with the exception of the latest pair. Cluster 3, which corresponds to the first issues of the Journal, is strongly characterized by expressions from the then current Psychology. Let us recall that, in this earlier phase, the Journal regularly published articles on psychological topics. Typical examples of titles from this cluster include: "The genetic method in psychology" (Washburn 1904), "The physiological argument against realism" (McGilvary 1907), and "The fitness of the environment for the continuity of consciousness" (Feingold 1914). The subsequent Cluster 1 is characterized by a massive presence of non-English (mostly German) words. These are mainly due to the titles of reviews of non-English monographs—chiefly of books published in continental Europe, such as "*L'Idea e il Mondo, Abbozo Introduttivo ad un Idealismo Concreto* by R. Pavese" (Boas 1926) and "*Das Psycho-*

Table 2.8 The clusters and their distinctive keywords

Cluster 3 (1904–1922). Genetic, psychologie, tests, physiological, nervous system, text-book, sensations, intelligence, elementary, consciousness, Spaulding, training, comparative psychology, Spencer, experimental study, holt, advertising, attention, psychical, laboratory, imagery, Bergson, mental, functional, pragmatism, über, experiments, psychological, affective, educational psychology, Pitkin, educational theory, energy, experimental psychology, children, efficiency, experimental, applied, relation, secondary, fundamental, über, essentials, child, Perry, visual, animal, outline, phenomena, suggestion, group, complex, behaviour, formal logic, practice, definition of consciousness, school, organisms, new realism, mental tests, nature of consciousness, psychology
Cluster 1 (1923–1941, 1948). American philosophical association, eastern division, annual meeting, sociologie, indian, jahrhunderts, correspondence, wesen, logik, historique, intelligence, droit, opinions, esprit, scientifique, survey, pensée, mathematik, meeting, contemporaine, rôle, études, metaphysik, untersuchungen, century, développement, contemporary german philosophy, Aristote, Platon, wirklichkeit, dieu, comte, lidée, international congress of philosophy, western division, positivism, métaphysique, modern, symbolism, erkenntnis, française, kritik
Cluster 2 (1942–1947, 1949–1962). Existentialism, vida, emotive, philosophical analysis, mead, quest, existentialist, Stevenson, Kierkegaard, myth, social sciences, american, época, vagueness, obra, influencia, democratic, synthetic, buddhism, moral judgments, uses, conference, humanities, political philosophy, questions, men, filosofía, value judgments.
Cluster 5 (1963–1991, 1993, 1995, 1999). Identity, rationality, quine, Scheffler, thesis, indeterminacy, Rawls, autonomy, revisited, utilitarianism, reduction, reference, frege, explanations, rules, reasons, goodman, acts, scepticism, epistemic, Katz, Levi, moral dilemmas, Anthony Kenny, actions, modality, wittgenstein, utility, metaphor, transcendental, decision, ontological, persons, explanation, morality, Williams, discourse, Daniel, violence, salmon, Hacking, justice, Kripke, foundationalism, intentionality, semantics, intention, responsibility, identity theory, models, property, punishment, argument, causal, consistency, beliefs, moral, understanding, egoism, inductive logic, Rosenberg, later, conditional, have, paradoxes, justified, extensional, preference, sexual, Nelson, philosophical papers, arithmetic, distributive justice, interpersonal, probabilities, essential, abortion, naturalized, Martin, counterfactuals, analytical, emotions, modal, intentional, possibility, reflections, virtue, arguments, agency, perceptual
Cluster 4 (1992, 1994, 1996–1998, 2000–2016). Physicalism, egalitarianism, puzzle, epistemic, memoriam, lecture, luck, conditionals, multiple, argument, games, colour, kind, counterpart theory, realization, evolutionary, context, Gödel, causation, reasons, pain, knowing, decision theory, alternative, possibilities, belief, pluralism, vague, solving, parts, self-knowledge, conditionalization, rationality, justice, agency, perception, control, worlds, rule, theory, properties, Thomson, impossibility, sex, scientific realism, libertarianism, debate, extended, game, objects, metaphysics, probability, explanatory, moral responsibility, lying, Nagel, scope, Bayesian, supervenience

Physische Problem by Robert Reininger" (Hausheer 1931). Cluster 2 is the latest one involving non-English words, and it is characterized by a significant presence of pragmatist and existentialist philosophers and topics. Examples of titles from this cluster are: "Fetishism in the existentialism of Sartre" (Ames 1950), "Phenomenological method from the standpoint of the empiricistic bias" (Winthrop 1949), and "Value judgments, emotive meaning, and attitudes" (Ladd 1949). For our purposes, the most noticeable feature of the latest two clusters is that they involve a considerable number of authors and themes from analytic philosophy. Typical examples of titles from these clusters are "The logic of the identity theory" (Brandt and Kim 1967), "Indeterminacy of translation and theory" (Humphries

1970), "Physicalism and the Necessary a Posteriori" (Stoljar 2000), and "A puzzle about disbelief" (Ostertag 2005).

2.3.2 *Full-Text Corpus*

Topic extraction (see Chap. 10) is based on Reinert's method implemented by Iramuteq software package (Ratinaud and Marchand 2012). Such method consists in classifying small portions of text into classes or topics, also called "lexical worlds" (Reinert 1983, 1990). A topic is a specific vocabulary—that is, a set of words—whose properties depend, intuitively, on the subject matter of the underlying text. More specifically, topics are classes of words that are more likely to occur together in the same text-chunks of the corpus.

Topics are obtained from a corpus in the absence of any prior information. The process of topic extraction may be described as follows. First, Reinert's algorithm divides certain portions of text (full articles, in our case) into units (Elementary Units of Context, ECUs), viz., chunks of similar length. These chunks may be sentences—or fragments thereof—that are demarcated by punctuation marks. Second, the algorithm individuates, within each unit, the occurrences and co-occurrences of lexical words (i.e. words endowed with content such as names, verbs, or adjectives, as opposed to grammatical words such as articles or conjunctions, which are ignored). The outcomes of this preliminary survey are put into a matrix (words × units), and the level of similarity between units is evaluated on the basis of the occurrence or co-occurrence of words within units. The result is a *dendrogram*, that is, a hierarchical, tree-like diagram. Based on this diagram, the algorithm individuates the clusters and the factors which better represent a specific topic (classes of units involving words that are relevant for the same cluster). By classifying units in this way, it is possible to automatically select only portions of texts concerning the same topic. Moreover, for each cluster, the list of the most significant words can be identified (Reinert 1999).[6]

Topic extraction allows to get information about the content of a corpus in a very quick way. It provides an automatic classification of the text within the corpus into clusters, which reflects its main areas of content. Being based on completely mechanical and unsupervised procedure, its results should not be affected by biases from the analyst.

By using topic extraction, we singled out 12 topics. Topics contain the most relevant words of the ECUs classified inside them. As a whole, the topics account for 97.07% of the ECUs (193,606 ECUs out of 199,447). These topics, which are organized into three broad groups, can be made to correspond to both thematic distinctions and functional distinctions.[7] We decided to give them synthetic descriptions

[6] For a more detailed introduction to topic analysis, see Sbalchiero and Tuzzi (2016).

[7] By a thematic distinction, we mean a distinction concerning the theme a given form is related to. For instance, "art", "aesthetic", and "poetry" are all words related to aesthetically relevant topics.

that, in our mind and according to other expert consultants, reflect the corresponding theme or function. The percentage following the description of a class expresses the relative contribution of the class to the whole corpus. It must be noted that a unique form can occur in more than one topic (Table 2.9).

It is worth stressing that these classes were selected on purely statistical grounds, with no involvement of human experts since it is based on an unsupervised classification algorithm. We surmise that the automatic topic extraction software did a very good job in selecting classes that are very easy to be made to correspond to recognizable topics, and to topics that cover the main thematic distinctions within the field of philosophy.

In what follows, we shall ignore classes 11 and 12 (and, thus, group 3) as broadly irrelevant to our main concerns. As per groups 1 and 2, they are very variegated, and it is not easy to spot a significant silver thread within them. In a very generic fashion, however, we can say that group 2 is dominated by practical philosophy and by topics characteristic of "continental" metaphysics. In contrast, group 1 is more theoretical in character, with a preeminence of scientific, logico-epistemic, and linguistic concerns.

If any process of rise to dominance of analytic philosophy occurred between the post-Second war and the 1970s, then it is reasonable to expect that topic 7 (*Continental metaphysics, existentialism, spirituality*) gradually fell in importance in that period, in favour of distinctively "analytic" topics such as semantics, logic, and epistemology. This expectation is indeed confirmed by our data.

Figure 2.3 illustrates the way in which our ten topics of interest distribute over the 1946–1975 period: intuitively, darker cells correspond to years when forms belonging to the topic are most likely to be found. Thus, for instance, topics 2, 3, and 7 play a much more important role in earlier volumes (up to the early 1960s) than in later ones, and vice versa for topics 4, 5, and 10. The general pattern is similar to the one seen when dealing with the title corpus: the mid-1960s marked an important lexical turn in the Journal. We remain neutral as to whether this transformation reflects a change in the themes the authors were dealing with or in their methods of inquiry. In any event, it appears that, during the 1940s and the 1950s, the articles published in the Journal were closer to "continental" philosophical strands, and more concerned with aesthetic topics and with broadly cultural, social, and political issues than in the subsequent decades. Significantly, however, the topics corresponding to ethics does not display the same decreasing pattern. Thus, it appears that the lexical change within the Journal cannot be described simply as a turn from practical to theoretical interests.

By a functional distinction, we mean a distinction concerning the role of forms within the *corpus*. For instance, in the Journal, words for concrete actions and things such as like "kill", "shoot", "eye", and "arm" are typically used to offer concrete examples.

Table 2.9 Dendogram of the Journal (full-text) topic. For each topic, some of the words shown to be most characteristic by the chi-square test are listed. All words listed have a p value <0.0001

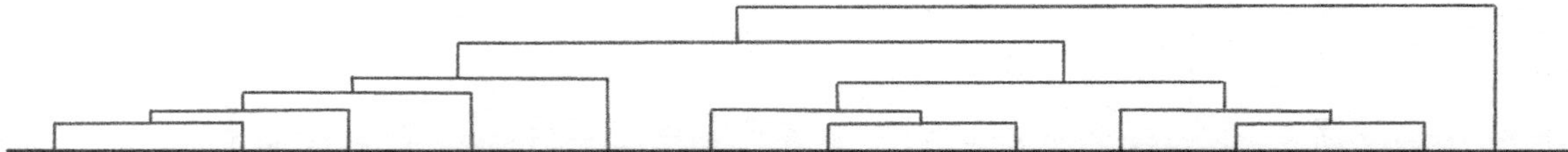

Class 3	Class 2	Class 7	Class 9	Class 11	Class 6	Class 5	Class 4	Class 10	Class 8	Class 1	Class 12
11.5%	2.7%	10.2%	9.1%	8.5%	3.4%	7.3%	10.8%	10.1%	9.8%	13.7%	2.9%
Life, culture, society, politics	*Aesthetics*	*Continental metaphysics, existentialism, spirituality*	*Ethics*	*Concrete examples*	*Physics, empirical science*	*Math, set theory, model theory*	*General semantics*	*Logic*	*Epistemology*	*Argumentative prose, philosophy of language*	*Academic meetings and publication details*
life	art	God	act	kill	space	set	property	true	hypothesis	question	university
activity	work	Heidegger	right	shoot	particle	member	object	false	evidence	distinction	press
process	artist	philosopher	action	eye	metric	numb	predicate	statement	conclusion	answer	new
experience	aesthetic	religion	duty	arm	velocity	k	name	proposition	inference	what	symposium
social	beauty	religious	morally	hear	Newton	sequence	relation	sentence	argument	analysis	American
environment	artistic	faith	ought	foot	Einstein	axiom	entity	truth	induction	language	philosophical
organism	aesthetic	metaphysics	obligation	water	geometry	input	class	assert	principle	mean	association
control	experience	philosophy	good	house	electron	model	attribute	entail	probability	between	journal
nature	critic	existentialism	punishment	car	time	output	designate	falsity	justification	problem	London
cultural	appreciation	man	wrong	eat	temperature	set_theory	term	believe	support	Wittgenstein	reprint
man	poetry	Plato	moral	pull	motion	number	denote	conjunction	justify	clear	Columbia
growth	Croce	Hegel	legal	fire	measurement	variable	thing	conditional	inductive	word	Eastern
achieve	beautiful	world	agent	tree	physics	theorem	event	antecedent	accept	how	division
end	fine	thinker	government	king	simultaneity	finite	identity	assertion	knowledge	criterion	annual
human	quality	Sartre	desire	night	quantum	formula	noun	imply	establish	definition	hall
value	aesthetically	spirit	utilitarian	pain	Euclidean	function	verb	utterance	confirmation	discuss	Harvard
live	imagination	Christian	justice	hour	atom	infinite	refer	equivalent	probable	linguistic	Oxford
development	sublime	existentialism	morality	sea	molecule	subset	exist	negation	deductive	theory	MacMillan
capacity	enjoyment	idealism	bad	Jones	interval	system	temporal	contradictory	premise	difficulty	Chicago
goal	beholder	divine	punish	clock	Newtonian	assign	universal	express	certainty	notion	Prentice
economic	spectator	Kierkegaard	motive	air	mechanic	outcome	substance	utter	reason	suggest	California
organization	style	tradition	judge	death	physicist	machine	particular	premise	test	attempt	Englewood
education	poetic	phenomenology	law	star	Lorentz	integer	singular	consequent	deduction	ordinary_	cliffs
own	art object	love	promise	finger	gas	sentence	singular_	speaker	rule	language	proceeding
past	medium	soul	conscience	sun	hydrogen	arithmetic	term	belief	claim	term	edit
intelligence	aesthetic	modern	person	paint	relativistic	class	distinct	contingent	valid	meaning	Harper
ideal	value	Nietzsche	excuse	fish	equation	natural	occurrence	denial	memory	usage	publish
organize	aesthetician	Husserl	doing	die	momentum	number	symbol	falsehood	validity	propose	Aristotelian
tension	externalizati	theologian	policy	walk	geometrical	pair	possible	necessarily_	proof	point	society
consciousness	on	doctrine	citizen	watch	oxygen	notation	world	true	basis	account	philosophy
emotional	comic	Spinoza	society	ball	geodesic	automaton	adjective	contradiction	prediction	clarify_	of science
society	admire	Descartes	crime	color	handedness	map	complex	disjunction	Hume	raise	Schilpp
individual	sensibility	eternal	evil	wall	curve	let	sign	contingency	Inductive_	view	essay
	music					Skolem	define	truth_value	inference	objection	page
	contemplation						proper_				references
	poem						name				paper

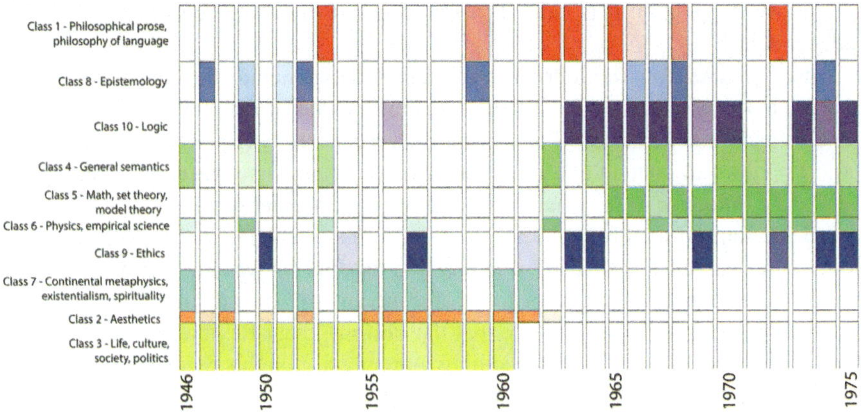

Fig. 2.3 Over-representation of topics in the time span 1946–1975. The height of the lines of each class is proportional to the dimension of the class in terms of the number of ECUs it contains. The width of the cells is proportional to the frequency of the ECUs in a given year. The tone of the colour is proportional to the strength of the association between class and year.

2.4 Some Conclusions About the Analytic Turn

There are two philosophical questions that one can ask at this point. The first one is philosophical in the sense that it has philosophy (or history thereof) as its subject matter: what conclusions can we draw about US analytic philosophy based on the data discussed so far? The second question is instead philosophical in nature: how does the statistical analysis of lexical data support conclusions about the history of philosophy? In this section, we focus on the former question, which we have already tackled in the introduction, and in the next (and final) section we shall deal with the latter one.

To begin with, the above data support a few conclusions that concern the Journal. Let us focus specifically on the contrast between papers published during the 1940s and the 1950s ("early sub-corpus") and papers published later, up to 1975 ("late sub-corpus"). Apparently, the early sub-corpus exemplifies a wider range of philosophical approaches than the late one, including existentialism, pragmatism, and phenomenology. Besides, the articles within the early sub-corpus are more sensitive to foreign, "continental" and, more generally, non-British influences. From a thematic viewpoint, the papers in the late sub-corpus are less prone to deal with political or social issues. Vocabulary from logics, mathematics, and argumentative prose are more likely to occur in the late sub-corpus. All in all, it appears that the years surrounding 1960 mark a turning point in this process of change, and that the late sub-corpus is dominated by typically analytic approaches and concerns. This pattern is confirmed by a recent qualitative study on the rise of analytic philosophy within the Journal (Katzav 2018). It is worth observing that the lexical turn within the full-text corpus coincided with a major change in the editors' panel of the journal (see above, Sect. 2.2). It is natural to suppose that these two facts are not wholly independent.[8]

Based on data from the Journal, one is tempted to conclude that analytic philosophy did not play a major role in the USA until the end of the 1950s. However, this conclusion would be hastened. For instance, a recent study (carried out by means of expert reviews of content and not by statistical methods) suggests that, during the 1950s, the influence of pre-analytic traditions (in particular pragmatism) was stronger in the Journal than in the other top US journal, the *Philosophical Review* (see Katzav and Vaesen 2017). Thus, it is possible the Journal was lagging behind other pieces of academic literature as to what concerns the overall impact of analytic philosophy—but it is also possible that the Philosophical Review was ahead of its time under this respect. To say something more precise on this point, a wider corpus would have to be taken into consideration. However, based both on our study and on the historical literature, it is reasonable to suppose that analytic philosophy was not dominant in the US philosophical scene until the 1950s—and possibly well into the 1950s (see, e.g., Putnam 1997; Soames 2008; Beaney 2013).

As mentioned in the introduction, this conclusion seems to run against a common notion among analytic philosophers—especially those not directly concerned with historical investigations. For instance, Føllesdal (1997, p. 1) has it that "analytic philosophy has dominated in the English-speaking world over the last sixty years [i.e., since the late Thirties]", whereas Stroll and Donnellan (2017) depicts analytic philosophy as "dominant in the Anglo-American world from the early 20th century".

2.5 More General Issues

Now let us turn to the second question mentioned above: how does the lexical analysis of philosophical journals based on statistical techniques support conclusions in the history of philosophy?

There are many things to be said about this question. Firstly, the statistical analysis provides essentially sets of words and distances between sets, which can be represented in geometrical space. For these data to be connected with historiographic hypotheses about content, the sets need to be associated with properties of their elements, the distances need to be checked by means of a survey of the content of these elements, and the geometrical representations need to be translated into qualitative descriptions. All this work is made based on previous knowledge about the topic and on general linguistic competence. For instance, to collect together *inference* and *conclusion* under the label "argumentative prose", semantic knowledge about these forms is needed, and such knowledge does not come from the statistical analysis. To make another example, we can regard a diachronic difference in the lexicon of a journal as relevant for the philosophical community only based on

[8] Katzav (2018) explicitly hypothesizes a connection between the composition of the editors' panel and the publication policy of the Journal.

certain empirical assumptions that, albeit seemingly obvious, are not part of the statistical set up—for instance that there was no abrupt change in the kind of philosophers that published in the journal during the period under consideration.

Secondly, the quantitative study of the philosophical lexicon can provide valuable information about the history of philosophy only based on previous knowledge from old-style historical sources and investigations. For instance, we concluded that the process of rise to dominance of analytic philosophy accompanied itself with a kind of lexical turn in philosophy because, among other things, we already knew that the analytic turn did occur during the period covered by the full-text corpus. Moreover, given that the lexical turn detected in our study complies with some natural expectations about the lexicon of analytic philosophy (for instance, a more-than-average frequency of technical terms from logic or semantics), the timeline of the lexical turn can be taken to provide information about the timeline of the analytic turn (at least insofar as the Journal is concerned). In other words, information about the lexical evolution of our corpora allows to collocate more precisely the analytic turn, but only given certain previous hypotheses and expert knowledge. This is a special example of a more general pattern: a balance between initial hypotheses or knowledge and subsequent refinements or revisions is common in many fields of inquiry.

Relatedly, the choice of the corpus is by no means accidental—rather, at least hopefully, it is driven by well-informed hypotheses. In the statistical methodology, general constraints have been formulated regarding the corpus to be examined. Two important requisites, partially in conflict with each other, are an adequate size of the corpus and a certain level of homogeneity of the texts within the corpus. A problematic case arises when homogeneity is weak and boils down to a very generic sort of similarity. For cases of this kind, it has been held that the corpus should include all the texts of the type under investigation or at least must be as broad as possible. For example, it has been argued that only by taking into consideration the totality of the literary production of a certain period it is possible to grasp some relevant characteristics of that production. This is the central and crucial part of the Distant Reading programme, due to Franco Moretti, which—metaphorically—consists of looking at a lot, possibly everything, from above and from far away (see Moretti 2013). Such kind of "reading", made possible by the use of statistical and computer tools, allows to highlight aspects that are not within the reach of single scholars, who are inevitably bound to consider only a small number of texts.

It is worth recalling that the need to widen the scope of the gaze had already been established, albeit with different perspectives and modalities, in the context of some historical studies. In Italy, Eugenio Garin promoted and practised a history of philosophy based on the following general ideas: (1) sometimes philosophy is also present in literary and scientific texts (as it happened, according to Garin, during Italian Humanism), and (2) the presence of philosophy, as well as its characteristics, can be adequately detected only considering a variety of works, sometimes very large and not limited to the texts of the "major" authors (see, e.g., Garin 1965, 1969, 2008).

The need to consider very large corpora is also implicit in the programme of the history of ideas, according to which an idea should be followed and traced over time through the expressions that have been used to expose and develop it. The programme was formulated and initiated about a century ago by Arthur Oncken Lovejoy (1873–1962) through his educational-cultural activity and the work *The Great Chain of Being: A Study of the History of an Idea* (Lovejoy 1936). Characteristic of this programme are the conception of ideas as unit-ideas and the interdisciplinary character of research. Unit-ideas are basic components of complex systems and can remain the same even if they become part of different systems. The procedure of the history of ideas is said to be "somewhat analogous to that of analytical chemistry" (p. 3).

It has been emphasized that Lovejoy talks about ideas as if they were invariant units, apparently atomistic in nature, which give rise to different systems through their different combinations. Such a perspective would ignore the fact that ideas change and, perhaps more importantly, that their connections can change, too. If the history of ideas is done taking into the account linguistic forms, there is also the obvious difficulty that different ideas may be signified by the same form and a same idea can be expressed by different forms. Prima facie, a statistical analysis, which tries to detect contents and their changes through data concerning linguistic forms, faces similar difficulties. If we assume that there are basic linguistic items, we can object that their meanings can be different in different texts, that they can change over time even in texts by the same author, and that these meanings can be differently expressed both in different texts and in the same text. Furthermore, no less difficulties are involved in identifying complex linguistic items or linguistic items having complex meanings. And all this without counting the issues of indeterminacy and vagueness and the issues of the relations between the meanings of the linguistic items that cannot be ignored when the target is the identification of contents.

There is an answer to these obvious objections. The purpose of a statistical-mathematical survey on content by means of computational tools is not, and does not presuppose, the identification of the contents of individual products, but rather the identification of themes and thematic modifications in very large corpora based on suitably defined similarity relationships. The identification of themes and thematic modifications in very large corpora need not require a very precise determination of topics and concepts and can neglect limited meaning oscillations or shifts. Moreover, if well done, a statistical analysis is less influenced by the knowledge and inclinations of the researcher in the collection of data.

A more specific issue concerning our research about the analytic turn arises from a feature that is quite often attributed to the analytic philosophy, i.e. that of being characterized for its methods rather than its contents. It is often said that analytic philosophers mainly look for arguments, evaluate arguments, and try to build arguments. That suggests that the analytic-philosophical orientation should be mainly manifested by the use of a language well suited for the analysis, the evaluation, and the building of arguments. Terms belonging to such a language are, for example, *valid* and its derivatives, *entail* and its derivatives, *conclude* and its derivatives, and so on. We effectively got confirmation of a limited growth over time of the use of such "logical" or "argumentative" vocabulary. However, it appears that the rise of analytic philosophy also manifested itself in the growth of interest towards certain themes as opposed to others.

A final note. Coherently with the purpose of this book, we limited ourselves to a purely content-metric approach to our corpora. We believe, however, that it would be useful to integrate the data presented here both with scientometric data and with historiographic information about the Journal. For instance, it would be interesting to take into consideration the mentions of recognizably analytic philosophers within the full-text corpus, and to carefully analyze the philosophical orientations of the Journal's editors. We leave these integrations for future work.

References

Ames, V. M. (1950). Fetishism in the existentialism of Sartre. *Journal of Philosophy, 47*(14), 407–411.

Beaney, M. (2013). The historiography of analytic philosophy. In M. Beaney (Ed.), *The Oxford handbook of analytic philosophy* (pp. 30–60). Oxford: Oxford University Press.

Boas, G. (1926). Review of "L'Idea e il Mondo, Abbozo Introduttivo ad un Idealismo Concreto" by R. Pavese. *Journal of Philosophy, 23*(10), 277–279.

Brandt, R., & Kim, J. (1967). The logic of the identity theory. *Journal of Philosophy, 64*(17), 515–537.

Feingold, G. A. (1914). The fitness of the environment for the continuity of consciousness. *The Journal of Philosophy, Psychology and Scientific Methods, 11*(16), 436–441.

Føllesdal, D. (1997). Analytic philosophy: What is it and why should one engage in it? In H. Glock (Ed.), *The rise of analytic philosophy* (pp. 1–16). Oxford: Blackwell.

Garin, E. (1965). *Italian Humanism: Philosophy and Civic Life in the Renaissance (Trans: Munz, P.)*. New York: Harper & Row.

Garin, E. (1969). *Science and Civic Life in the Italian Renaissance (Trans: Munz, P.)*. New York: Doubleday.

Garin, E. (2008). *History of Italian Philosophy (Trans: Pinton, G.)*. Amsterdam/New York: Rodopi.

Greenacre, M. J. (2007). *Correspondence analysis in practice*. London: Chapman & Hall.

Hausheer, H. (1931). Review of "Das Psycho-Physische Problem" by Robert Reininger. *Journal of Philosophy, 28*(26), 716–719.

Humphries, B. (1970). Indeterminacy of translation and theory. *Journal of Philosophy, 67*(6), 167–178.

Katzav, J. (2018). Analytic philosophy, 1925–69: Emergence, management and nature. *British Journal for the History of Philosophy*, 1–24.

Katzav, J., & Vaesen, K. (2017). On the emergence of American analytic philosophy. *British Journal for the History of Philosophy, 25*(4), 772–798.

Ladd, J. (1949). Value judgments, emotive meaning, and attitudes. *Journal of Philosophy, 46*(5), 119–128.

Lebart, L., Morineau, A., & Warwick, K. M. (1984). *Multivariate descriptive statistical analysis. Correspondence analysis and related techniques for large matrices*. New York: Wiley.

Leiter, B. (2015). Top philosophy journals, without regard to area. *Leiter Reports: A Philosophy Blog*. Retrieved March 18, 2018 from http://leiterreports.typepad.com/blog/2013/07/top-philosophy-journals-without-regard-to-area.html.

Lovejoy, A. O. (1936). *The great chain of being: A study of the history of an idea*. Cambridge, MA: Harvard University Press.

McGilvary, E. B. (1907). The physiological argument against realism. *Journal of Philosophy, Psychology and Scientific Methods, 4*(22), 589–601.

Moretti, F. (2013). *Distant reading*. London: Verso Books.

Murtagh, F. (2005). *Correspondence analysis and data coding with java and R*. London: Chapman & Hall.

Ostertag, G. (2005). A puzzle about disbelief. *Journal of Philosophy, 102*(11), 573–593.

Putnam, H. (1997). A half century of philosophy, viewed from within. *Daedalus, 126*(1), 175–208.

Ratinaud, P., Marchand, P. (2012). Application de la méthode ALCESTE à de "gros" corpus et stabilité des "mondes lexicaux": Analyse du "CableGate" avec IRaMuTeQ. In *Actes des 11eme Journeés internationales d'Analyse statistique des Données Textuelles* (pp. 835–844). Liège, Belgique.

Reinert, M. (1983). Une methode de classification descendante hierarchique: application a l'analyse lexicale par context. *Les Cahiers de l'Analyse des Donnees, 8*(2), 187–198.

Reinert, M. (1990). Alceste, une méthodologie d'analyse des données textuelles et une application: Aurélia de Gérard de Nerval. *Bulletin de Méthodologie Sociologique, 26*, 24–54.

Reinert, M. (1999). Quelques interrogations à propos de l'"objet" d'une analyse de discours de type statistique et de la réponse "Alceste". *Langage et Societe, 90*, 57–70.

Sbalchiero, S., & Tuzzi, A. (2016). Scientists' spirituality in scientists' words. Assessing and enriching the results of a qualitative analysis of in-depth interviews by means of quantitative approaches. *Quality and Quantity, 50*, 1333–1348.

Soames, S. (2008). Analytic philosophy in America. In C. Misak (Ed.), *The Oxford handbook of American philosophy* (pp. 449–481). Oxford: Oxford University Press.

Stoljar, D. (2000). Physicalism and the necessary a posteriori. *Journal of Philosophy, 97*(1), 33–54.

Stroll, A., & Donnellan, K. S. (2017). Analytic philosophy. In *Encyclopædia Britannica*, Encyclopaedia Britannica (online). Retrieved March 21, 2018 from https://www.britannica.com/topic/analytic-philosophy

Washburn, M. F. (1904). The genetic method in psychology. *Journal of Philosophy, Psychology and Scientific Methods, 1*(18), 491–494.

Winthrop, H. (1949). Phenomenological method from the standpoint of the empiricistic bias. *Journal of Philosophy, 46*(3), 57–74.

Chapter 3
Exploring the History of American Sociology Through Topic Modelling

Giuseppe Giordan, Chantal Saint-Blancat, and Stefano Sbalchiero

Contents

3.1 Introduction 46
3.2 American Journal of Sociology: Corpus and Data 46
3.3 Topic Modelling: Results 47
3.4 Hot and Cold (*Sociological*) Topics 54
3.5 Conclusion: Sociologically Speaking, Where Do We Come from and Where Are We Going? 61
References 63

Abstract The study aims to explore the temporal evolution of topics in the abstracts of articles published by the *American Journal of Sociology* in the last century (1921–2016). Within the topic detection perspective, this study traces the topics with a significant increasing or decreasing trends and shows different shifts that involved the discipline in the American context. The relevance of a research area in a specific time span and its temporal evolution can be a way to identify the paradigm changes which show how sociologists have reacted to social phenomena and they have stimulated theories, ideas, and emerging research interests along with the social change it observed. The results highlighted the directions of sociological debate within the same period by differentiating the early period when the Journal was the voice of a new discipline—sociology—from recent developments within the constant tensions between specialization processes and the scientific diversification of the discipline.

Keywords American Journal of Sociology · Topic detection · Latent Dirichlet allocation · History of sociology

---

Authors are listed in alphabetical order, as they contributed equally to the present work.

G. Giordan · C. Saint-Blancat · S. Sbalchiero (✉)
University of Padova, Padova, Italy
e-mail: stefano.sbalchiero@unipd.it

A. Tuzzi (ed.), *Tracing the Life Cycle of Ideas in the Humanities and Social Sciences*, Quantitative Methods in the Humanities and Social Sciences,
https://doi.org/10.1007/978-3-319-97064-6_3

3.1 Introduction

We will begin this chapter with the following observation: sociology is a continuous approach to the understanding of current social changes and is by nature connected with the different contexts in which the discipline tooks shape. In early twentieth century American sociology, for example, Marx played a secondary role (Burawoy 1982) and "between the two world wars, the Chicago School was inspired by German thought as filtered through Louis Wirth and Robert Park. The two decades after World War Two were dominated by Talcott Parsons's grand synthesis of Weber, Durkheim, Pareto, Marshall, and, subsequently, Freud" (Burawoy 1982, p. 1). In the same period, European sociology considered Marx as a "classic" among the founders of modern sociology. And yet, the development of sociology in the Soviet Union was politicized after the Russian revolution to the point where it almost disappeared (Weinberg 1974); in China after the Communist revolution, sociology—considered a bourgeois discipline—was abolished from 1952 until 1979, when it was re-established (Zhou and Pei 1997). Indeed, what we know today as sociology is the result of theoretical thinking, a succession of new methodological directions and widespread internal debates that were brought into sharper focus during the discipline's development in different historical moments and in different contexts. During the twentieth century, then, sociology became increasingly complex and expanded its influence, thus leading to a diversification of the discipline. Although several studies have looked at the shape of the history of sociology (Kalekin-Fishman and Denis 2012; Turner 1998; Scott and Desfor Edles 2011), this chapter will focus on the intellectual grounding of American sociology, as suggested by the following consideration: "how rich the history of sociology is and how instructive exploring that history can be. If science aspires to transcend the limits of surface observation and prejudice, historical understanding of the conditions and trajectories of scientific work can be as valuable as theoretical frameworks and research methods" (Calhoun 2007, p. 38). Tracing the trends in sociology through topic modelling is part of the strategy of exploring how a discipline grows and evolves in time, along with the social change it observes. In such a framework, the choice of analysing the *American Journal of Sociology*'s evolution is anything but fortuitous, as it can legitimately be considered as a paradigmatic case study of the discipline's transformations from the late nineteenth to the twenty-first century.

3.2 American Journal of Sociology: Corpus and Data

The *American Journal of Sociology* (AJS), established in 1895 as the first US scholarly journal in its field, is considered America's preeminent journal for sociologists from all over the world. From its beginning until the launch of the *American Sociological Review* in 1936, the early issues of the AJS overlap with the disciplines' emergence (AJS was the de facto official journal of the American Sociological Association, or ASA), and in this period sociology was practically synonymous with the Chicago School. Apart from the years of the Depression and the New Deal when sociology lost prestige, influence, and jobs (Calhoun 2007, p. 29), AJS has covered all the historical stages of the discipline, from the Chicago period, to the post-World War II years, up to the Golden Age of the 1960s, the shift of the 1990s and the issues of our own day. Its ability to do so is largely due to the exceptional leadership of its successive editors: Small, Faris, Burgess, and Blumer, to name the most prominent, whose open-mindedness and foresight have made the difference.

The corpus used for this study includes all abstracts of the papers published in AJS. We decided to work with abstracts which provide concise information about the contents of published articles. We excluded material that did not provide information about the content, e.g., editorials, master heads, errata, acknowledgements, rejoinders, notes, announcements, corrections, lists of consultants, obituaries.

Overall, 122 volumes and 734 issues have been published up to 2016 (time span 1895–2016). The corpus consists of 3992 abstracts collected from Volume No. 27, Issue No. 1 (1921) to Volume No. 121, Issue No. 4 (2016) (mean: 41 per year), that is since the abstracts are available (Table 3.1).

During preprocessing, the corpus was tokenized and then normalized by replacing uppercase with lowercase letters. The corpus (Table 3.2) consists of 24,418 word-types (different words) and 512,410 word-tokens (occurrences). The number of hapaxes (i.e., types occurring only once in the corpus) is 11,495 corresponding to 47.08% of the forms in the corpus vocabulary (Table 3.2).

Through an automatic search procedure, we identified relevant Multiword Expressions (MWEs), i.e., informative sequences of words (see Chap. 8) repeated at least five times in the corpus. This criterion was based on the following consideration: if an MWE did not appear at least 5 times in the corpus, i.e., about once every 20 years, it was not considered important. A total of 849 MWEs were identified, of which 417 appeared with a number of occurrences equal to or greater than 10. This procedure recognizes each MWE in the abstract as a unique word (e.g., *social structure*, 115 occurrences).

It is worth looking at the excerpt of AJS vocabulary ordered by decreasing frequency (Table 3.3). Even if the frequency of a word is a rough indicator, it can be considered as a clue of the importance of certain topics in the corpus which correspond to the changes which took place in the structure of research activity: changes in theory and consequently in methodology and in the organization of research. The most common lexical forms (*study, social, economic, theory, individual, cultural, family, sociology, education, society, men, factors, changes, effect*) refer to a wide range of topics that characterized AJS publications, just as the most frequent MWEs (The United States, social structure, social relations) correspond to classics of sociology.

3.3 Topic Modelling: Results

As the AJS corpus is a valuable cross-section of American sociology, we decided to apply a topic detection procedure (Blei et al. 2003) to explore the main topics that have appeared in almost a century of publications. In this approach, a topic is characterized by a cluster of co-occurring words, and a document (in our case an

Table 3.1 Features of AJS corpus: volumes, issues, and available abstracts

Years	Volumes	Issues	Abstracts	*N*
1895–1900	1–6	33	0	0
1901–1905	6–11	30	0	0
1906–1910	11–16	30	0	0
1911–1915	16–21	30	0	0
1916–1920	21–26	30	0	0
1921–1925	26–31	30	138	22,853
1926–1930	31–36	30	261	45,521
1931–1935	36–41	30	271	39,895
1936–1940	41–46	30	193	30,737
1941–1945	46–51	30	242	31,419
1946–1950	51–56	30	227	19,728
1951–1955	56–61	30	232	19,807
1956–1960	61–66	30	249	23,040
1961–1965	66–71	30	207	21,696
1966–1970	71–76	30	219	26,837
1971–1975	76–81	30	244	29,776
1976–1980	81–86	30	218	29,876
1981–1985	86–91	30	204	28,962
1986–1990	91–96	30	181	26,767
1991–1995	96–101	30	183	22,367
1996–2000	101–106	30	169	19,896
2001–2005	106–111	30	158	20,125
2006–2010	111–116	30	187	23,805
2011–2016	116–122	36	209	29,303

Size in word-tokens (*N*)

Table 3.2 Lexical measures of AJS corpus

(*V*) Word-types	24,418
(*N*) Word-tokens	512,410
(*V*/*N*)*100 = type/token ratio	4.8
(*V*1/*V*)*100 = hapax percentage	47.1

abstract) is a mixture of these topics (Blei and Lafferty 2009; Grimmer and Stewart 2013). After applying topic detection, we investigated the main chronological shifts (Griffiths and Steyvers 2004) that have occurred in the Journal. This exploratory study not only revealed different types and shapes of sociological approaches over the years, but has also proved very useful in understanding "latent" research directions that are otherwise difficult to identify. Latent Dirichlet Allocation (LDA) analysis was implemented by means of the "topicmodels" package (Grün and Hornik 2011) available in R (R Development Core Team 2016) (see Chap. 10).

Preprocessing consisted in removing punctuation marks, numbers, and stop words (*the*, *if*, *and*, …), which are not usually used in topic modelling. To identify the number of topics that best fits the LDA model, we estimated the optimal number

Table 3.3 Excerpt of the lexical contingency table words × years

Words	Occurrences (corpus)	1921	1922	1923	:	1945	1946	:	2014	2015	2016
the	36,622	227	443	366	:	343	340	:	271	259	344
of	29,252	208	372	313	:	279	247	:	205	198	258
study	959	7	9	15	:	14	5	:	8	20	12
social	905	14	30	12	:	7	8	:	7	13	13
economic	648	4	13	6	:	4	6	:	6	0	4
theory	648	0	2	4	:	2	3	:	8	12	6
individual	512	5	16	4	:	3	6	:	6	4	1
cultural	510	3	12	5	:	2	7	:	10	8	5
family	502	0	3	0	:	11	9	:	8	7	3
sociology	469	9	14	13	:	12	4	:	1	2	2
state	464	8	12	3	:	1	0	:	6	5	10
education	464	4	11	14	:	1	1	:	1	1	2
society	427	2	14	5	:	10	4	:	1	2	0
men	426	0	3	2	:	2	6	:	2	2	8
factors	421	1	2	2	:	4	6	:	5	5	3
changes	418	0	4	2	:	2	3	:	3	4	7
effect	414	0	0	1	:	0	2	:	1	2	13
United States	395	0	2	3	:	3	4	:	4	1	6
children	388	0	3	4	:	2	5	:	5	5	6
culture	386	0	8	6	:	7	4	:	1	1	3
behavior	384	1	2	5	:	6	5	:	1	7	1
class	376	1	8	1	:	2	3	:	5	4	2
american	369	5	3	5	:	9	6	:	4	4	2
studies	368	0	1	0	:	3	3	:	3	6	3
development	367	2	5	5	:	7	2	:	2	3	1
social structure	115	0	0	1	:	2	0	:	1	0	1
social science	101	1	3	8	:	3	0	:	1	0	0
social relations	45	1	0	0	:	1	0	:	0	0	0
resourcefulness	1	0	0	0	:	0	0	:	0	0	0

of topics using the log-likelihood variation (Griffiths and Steyvers 2004) from 2 to 50 topics. The results suggest that the best number of topics is around 30 (Fig. 3.1).

We ran the Latent Dirichlet Allocation (LDA) using the "topicmodels" package (Grün and Hornik 2011). As can be seen from the most probable words for each topic (Table 3.4), the words in the same topic are semantically associated and refer to the same issue.

Even though only a few words are shown for each topic, it should be mentioned that the topics exemplify the complexity of American sociology during the last century, with its reflections on a vast range of issues affecting the public sphere. Using per-document topic probabilities, we can briefly interpret each topic: rural life and farm populations (topic 1); racial differences (topic 2); scientific investigation of migration (topic 3); relationship between delinquency and other variables (topic 4); social organization and rural vs. urban communities (topic 5); Weberian paradigms

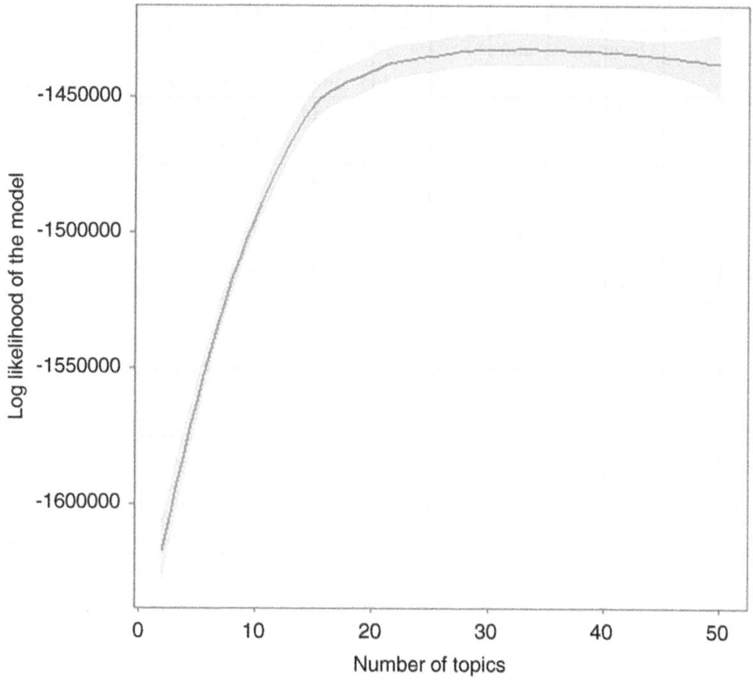

Fig. 3.1 Log-likelihood for increasing numbers of topics

in social science (topic 6); debate about sociology as a scientific discipline (topic 7); deviance and crime (topic 8); psychosocial interpretations of social processes (topic 9); military morale, civilian morale, and public opinion (topic 10); gender (topic 11); classes and cultural differences (topic 12); the racial question (topic 13); transformations of urban and suburban cultures (topic 14); power and élite (topic 15); study of Asian social groups (topic 16); mobility and occupational mobility (topic 17); sociology of religion and social function of religion (topic 18); attitude and prejudice (topic 19); market and social structure (topic 20); government and health education (topic 21); social influence (topic 22); female participation and occupational prestige (topic 23); marriage and school completion (topic 24); ethnic groups and political action (topic 25); industries in the urban economy (topic 26); protest, democracy and violence (topic 27); use of statistics and predictive modelling (topic 28); family structure (topic 29); real and perceived attitude (topic 30). Before going into detail on the content, which would deserve a wider interpretative effort, the topic detection analysis shows how American sociology developed priorities in the social research agenda by specializing and fostering the birth of a range of sub-disciplines.

Albion Small was both a pioneer and a visionary. He had two clear and ambitious goals: to establish sociology as an autonomous discipline, and to maintain a strong link between theoretical analysis and the unlimited variety of social research. As he

Table 3.4 Excerpt of five most probable words for each topic (decreasing order of probability)

Topic 1	Topic 2	Topic 3	Topic 4	Topic 5	Topic 6
world	racial	cent	variables	community	weber
countries	black	rate	delinquency	communities	identity
labor	white	rates	measures	movement	professional
farm	race	migration	association	immigrant	authority
farmers	whites	urban	income	leadership	identities
Topic 7	**Topic 8**	**Topic 9**	**Topic 10**	**Topic 11**	**Topic 12**
sociology	crime	personality	war	women	class
science	criminal	behavior	morale	gender	city
scientific	rates	person	military	men	middle
sociologists	police	emotional	public opinion	organizational	pattern
knowledge	crimes	man	public	organizations	urban
Topic 13	**Topic 14**	**Topic 15**	**Topic 16**	**Topic 17**	**Topic 18**
culture	agreement	power	concept	occupational	church
negro	suburbs	network	chinese	occupations	mental
conflict	mead	exchange	japanese	education	freud
leisure	suburban	networks	catholics	mobility	religious
union	property	elites	neighborhood	status	progress
Topic 19	**Topic 20**	**Topic 21**	**Topic 22**	**Topic 23**	**Topic 24**
attitudes	method	health	behavior	prestige	family
prejudice	market	government	norms	college	students
attitude	adjustment	education	trust	sex	marriage
revolution	markets	year	conformity	females	school
immigrants	logical	federal	status	expectations	marital
Topic 25	**Topic 26**	**Topic 27**	**Topic 28**	**Topic 29**	**Topic 30**
action	industrial	violence	models	children	party
ethnic	production	protest	variables	family	network
jewish	industry	democratic	techniques	families	friendship
politics	workers	author	prediction	child	attitude
jews	urban	law	statistical	parents	homophily

wrote in 1895, AJS "will be devoted to the organization of knowledge pertaining to the relations of men in society into a sociology that shall represent the best American scholarship. On the other hand, the Journal will attempt to translate sociology into the language of ordinary life, so that it will not appear to be merely a classification and explanation of fossil facts" (Small 1895, p. 13–14). This choice can still be seen today in the diversity of topics reinforced by the flourishing of specialist fields, the relevance of multidisciplinary approaches, and the impact of research on social reform. These are the two contrasting conceptions: building an independent intellectual enterprise, and remembering that the discipline's social and institutional function is also to grasp what is happening "out there." Their respective influence will alternate, with a focus on theory and sociology's identity during the formation of the canon in the Golden Age of the 1960s, and greater openness to social problems in the 1970s (civil rights movements, feminism, and racial issues).

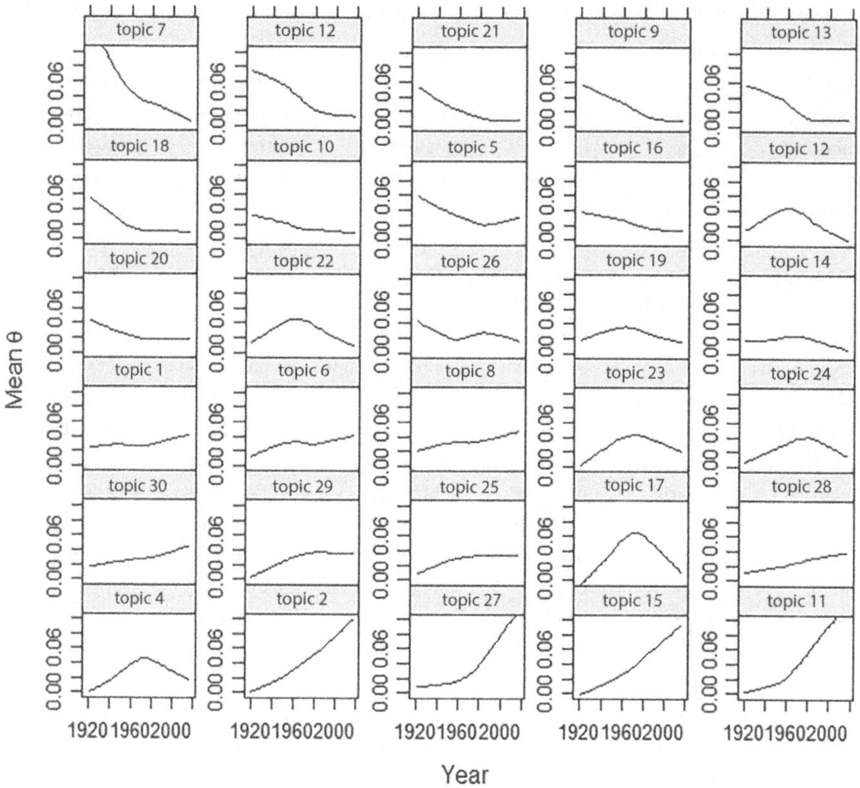

Fig. 3.2 Temporal patterns of topics in American sociology (1921–2016)

For this study, we identified the temporal evolution of topics by means of a linear trend analysis (Griffiths and Steyvers 2004) starting from the 30 topics of the fitted model. The temporal trends show the direction of significant shifts and variations in American sociology over the years (Fig. 3.2).

A look at the graphs for the most probable terms for each topic (Table 3.2) suggests a number of considerations. Since the topics show different trends and embrace theoretical and methodological shifts, we can first note that certain topics have grown over time, either from a methodological standpoint, by developing innovative statistical instruments for addressing the social issue concerned (topic 28), or in terms of their content, with a growing attention to emerging social issues during the period in question. This, for example, is the case of studies of the relationship between delinquency and certain social factors (topic 4), whose growth is in part similar to studies of deviance and crime (topic 8). Other increasing topics reflect certain issues that have emerged over the years: studies of collective protest and social processes among mobilized groups (topic 27) have increased since the 1960s and entered the sociological agenda, much like gender (topic 11), the debate on racial differences (topic 2), and power in exchange networks and power distribution

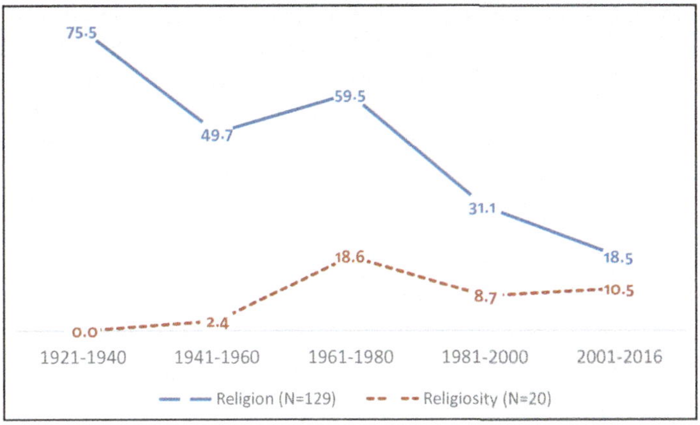

Fig. 3.3 Occurrences of religion and religiosity. Rates × 10,000 occurrences (The rates have been obtained as follows: No. occurrences of the word/No. of occurrences of the period * 10,000)

(topic 15). These topics indicate that sociology, and especially the themes of sociology as social criticism of society (social movements, social action, political power, feminist revolution, race) have had a significant influence on the development of American sociology since the 1960s. Secondly, we can see other topics whose trajectory has decreased: the sociological study of religion, the psychological approach to interpretations of social processes (topic 9), the racial question (topic 13), and the need to affirm sociology as a scientific discipline (topic 7).

It is particularly interesting that religion (topic 18) and the sociological approach to religion has decreased along with the need to establish sociology as a discipline, which characterized articles until the mid-1980s. Without going into the details of the contents, which would deserve further investigation, it is also worth mentioning how the topic religion has decreased over time (Fig. 3.3).

Taking into consideration only the form *religion* and *religiosity*, in fact, not only they are decreasing, according to the topic 18, but they are present in the corpus until the 1980s (Fig. 3.3). On one side, this suggests that the journal has assumed a more defined identity, for example distancing itself from other disciplines such as psychology, while on the other hand the social study of religion has established itself as an autonomous discipline.

Lastly, we can see other topics that were meteoric in their rise and near disappearance: mobility and occupational mobility (topic 17) were very popular in the 1970s and 1980s, as were female participation and occupational prestige (topic 23) and studies on family, marriage, and school completion (topic 24). Social influence and social control (topic 22), related to role behavior, status, values, and attitude and prejudice (topic 19) were popular until the 1950s and 1960s, when behaviorism and what we can call the behavioral revolution in sociology pushed them to the side. These changes in topics through the years show the discipline's vitality and hence its confident identity. They are also a measure of how American sociology's vigorous professionalism continues to evolve to reflect crucial issues in the national social agenda: the interest in race problems, criminality, statistical studies of population, and social inequalities have not lost relevance, while the study of immigrants is not central as it was in the days of the Chicago School. Likewise, the internal debate between the quantitative and qualitative approach is still as much at the core of the discipline as is the heterogeneity of its research fields.

3.4 Hot and Cold (*Sociological*) Topics

To focus on the major increases or decreases of topics from 1921 to 2016, i.e., on the topics whose trajectories rose or fell significantly, we analysed the coldest and hottest topics, providing the top term and abstract significantly correlated with these topics. The results enable us to restrict the field by analysing the five topics that rose or fell consistently in popularity by means of the significance level of the linear trend test statistic (p-level ≤ 0.0001). The coldest topics (Fig. 3.4) were very popular around the 1920s and 1950s, while hot topics indicate sociological developments from the 1960s and reflect how American sociology reacted to new social concerns.

The groups of coldest topics found through this analysis corresponded to certain specific theoretical perspectives and objects of research which decreased during the period considered. Developments in the discipline are apparent from a look at the most probable terms in the five coldest topics (Table 3.5).

First, we see the early debate on the institutionalization of sociology as scientific and academic disciplines with defined theoretical as well as methodological perspectives. This topic characterized the early debate in the journal (topic 7), and the problem was that of legitimizing knowledge production as "scientific." Some examples of the abstracts assigned to these topics, i.e., through high per-abstract topic probabilities, can be useful in understanding which abstracts pertain to topic 7. The abstract for the article by Florian Znaniecki entitled "The Object Matter of Sociology" is assigned to this topic with highest level of probability (99.2%):

> "*Necessity of determining the object matter of sociology*. In the present chaos of different conflicting presuppositions and methods found in sociological textbooks and monographs, it is impossible either to reach a systematization of the results of sociological research or to plan a rational program of future studies without a reconsideration of the current conceptions of the object matter of our science. *Sociology as a humanistic science*. Sociology must have a certain class of data as its object matter, and these data must be such as to allow a rational body of knowledge to be constructed about them. (…) Since it is impossible to combine any knowledge about natural facts with any knowledge about cultural facts into one logical system, sociology must choose whether it should be exclusively a natural or exclusively a humanistic science. The main interest of sociologists has always been in data with the humanistic coefficient, and it should in future confine itself to such. *Criticism of sociology as science of concrete societies*. The oldest definition of sociology is that of a generalizing and explanatory science of society taken as a concrete collectivity of human beings in their total cultural life (…)" (Znaniecki 1927, p. 529).

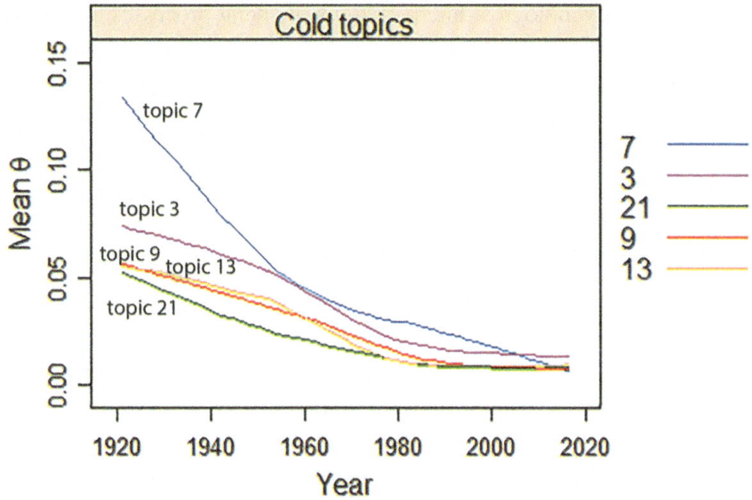

Fig. 3.4 Cold topics in *American Journal of Sociology*

Table 3.5 Top terms for cold topics (decreasing order of probability)

Cold topics				
Topic 7	Topic 3	Topic 21	Topic 9	Topic 13
sociology	cent	health	personality	culture
science	rate	government	behavior	conflict
scientific	rates	education	person	leisure
sociologists	migration	year	emotional	union
knowledge	urban	federal	man	white
concepts	born	medical	social psychology	class
social science	rural	legislation	learning	education
method	cities	administration	psychology	peoples
durkheim	fertility	states	adjustment	activities
phenomena	areas	public	institutions	family
concept	city	policy	experience	cultures
facts	decline	control	reactions	community
methods	age	programs	conception	institution
social phenomena	north	local	collective behavior	civilization
social sciences	proportion	law	happiness	world
sciences	south	agencies	reaction	societies
definition	migrants	relief	language	tradition
objective	birth	fields	stimuli	caste
techniques	native	program	interaction	competition
values	average	institutions	organism	emphasis

At the same time, psychological interpretations of social processes using a behavioristic approach (topic 9) decreased.

> "Mead considers that the data for solving the problems of social psychology were to be found in the conduct and experience of men rather than in the behavior of lower animals or in the facts of physiology. The ongoing social process was his starting-point, and man was assumed to be a part of nature, entirely and without residue. In the analysis of conscious reflective action he found use for all the psychological concepts. The social nature of man was assumed, and the self is shown to have its origin in communication which leads in man to self-stimulation and self-response and the taking of the role of the other. Personality is a role in a social situation. Every self is in a social context. (…) In his doctrine of 'the I and the me,' he found for himself an escape from a mechanistic view of human nature and a fresh defense of responsibility and freedom" (Faris 1937, p. 391).

The debate also increased on the "measurement" of certain social phenomena (topic 3) such as the migrants moving from rural to urban areas, with all the social urges they brought and the sociological reflections that accordingly emerged. Here, an example is provided by "Primary and Secondary Aspects of Interstate Migrations":

> "Writers on population movement have inferred that increase, in one state, of natives from another state is due to direct migration from the native state and that increase in this native state of natives from other states in an adequate measure of internal migration. (…) The fourfold aspect of this problem of interstate migration is presented graphically in the case of South Carolina. Indexes of migration have been developed for purposes of comparison, and it was found that while the rate of primary migration during the decade 1900-1910 was much lower for North Carolina- than for South Carolina-born Negroes, North Carolina-born Negroes who had moved from that state prior to 1900 were shifting more rapidly than were similar groups from South Carolina. Further, during 1910-20 there was a marked acceleration for South Carolina of primary, secondary, and total migration over the decade 1900-1910" (Ross and Truxal 1931, p. 435).

The emergence of a social reflection on the relationship between state legal policy and health care reform (topic 21) characterizes the debate on health and illness from the 1920s to the 1960s, with a focus on efforts to educate the public and to improve health legislation. The abstract for the article "Sociology Applied in the Field of Health" has the highest level of probability for this topic (99.4%):

> "There is at present a great increase in efforts for health. *Better educated public. Efforts of leaders*. The medical profession in not the only one responsible for securing better health. *Social and economic forces, correlation of those working toward same ideal from different view points*. The function of the physician is in the medical field, telling individuals and groups what is necessary for health. The attaining of these conditions is often not a medical matter. *Relation to such a condition as poverty to health*. The attaining of the conditions of health depends on individuals, groups, society as a whole, and on social agencies. (…) Lack of correlation of our knowledge in diverse fields is responsible for our lack of health. Medicine has received much of the blame. But the problem of producing health is to be solved only by concerted work in many fields" (Meredith 1922, p. 319).

Lastly, class culture, conflict, and leisure (topic 13) were popular issues in the 1930s and 1950s, when industrialization processes raised many questions that provided new insights for sociologists, as Alfred H. Lloyd emphasized in the article "Ages of Leisure" (99.5% per-abstract topic probability).

> "*Leisure as much a problem as work.* Not less important than the problem of labor and efficiency, of occupational expertness, is the problem of leisure, today pressing for reckoning. For many leisure is only rest or entertainment, but it is important to life constructively. Beset with dangers, it still is of evolutional value, enabling education, imagination, progressive adventure. *Three eras of leisure.* Witness the leisure afforded by long infancy and youth, in some sense the basis of man's superiority. Witness, again, the leisure and culture made possible by slavery domestic and proprietary or socially institutional. Witness, thirdly, the leisure which has been coming to man through automatic machinery, the 'Iron Man,' and which offers such adventure, such mixed danger and opportunity, as history has never known. *The new leisure and industrial democracy.* With this third leisure must be associated the call for an industrial democracy. All democracy would free men from some subjection and insure to all in some measure and to many in large measure the leisure attending the liberation and also the opportunity and importance attending the leisure; and, as to democracy today, with so much automatic and dehumanized instrumentation of life, leisure might properly even be added as a fourth natural right to life, liberty, and pursuit of happiness. (…) Thus the challenge of today comes from the Iron Man, the Giant Automaton, and no former culture can answer. Analogously, Virgil may have accompanied Dante, but he could not have replaced him" (Lloyd 1922, p. 160).

The cold topics indicate that AJS gradually moved away from social history, theoretical articles on social psychology, which were at the core of the study of personality, papers on social reform which reflect the Chicago sociologists' engagement in social problems, and the spirit of social reformism where social science should serve as a means of improving laws and policies. Among the founders, early sociology was influenced by Christianity, a characteristic which disappears around 1900. These shifts reflect a rethinking of the discipline's social function, which leaves room for a more secular approach to social issues, while still maintaining a focus on finding solutions to them and developing new fields of research, a strategy which will give rise to a number of sub-disciplines. After a decade, Small—far-seeing as usual—identified two lines of future development: one dedicated to theoretical research around "the whole social process," the other focusing on "applied sociology" (Small 1905, p. 2), thus heralding the period of Parsons, the discipline's second key catalyst (Calhoun 2007, p. 20). Shanas (1945), in his paper on the 50 years of AJS, underlines that the proportion of space devoted to statistical papers in population and methodology was to remain about 10% until the mid-1920s, when it rose sharply (Shanas 1945). This trend continued into the future. Thus, the impressive volume L, number 6 of May 1945 dedicated to the first 50 years of *AJS* not only provides a clear idea of the state of the art in the discipline, with outstanding scholars discussing the achievements of their field of expertise (Merton on Sociological Theory, E.W. Burgess on Social Research Methods, L. Wirth on Human Ecology, to name just a few), but also contains a paper on the proximate future of sociology by Florian Znaniecki (1945) entitled "Controversies in Doctrine and Methods," an issue we will find again in the hot topics (number 4).

The group of hottest topics (Fig. 3.5) is related to articles that mark a shift during the 1960s that underscores the most significant changes that have occurred.

Those years gave sociologists an opportunity to deal with social effervescence in a particular historical moment and, consequently, to engage in wide-ranging empirical reflection that led to an increase in new sub-disciplines and thus to new specializations (Table 3.6).

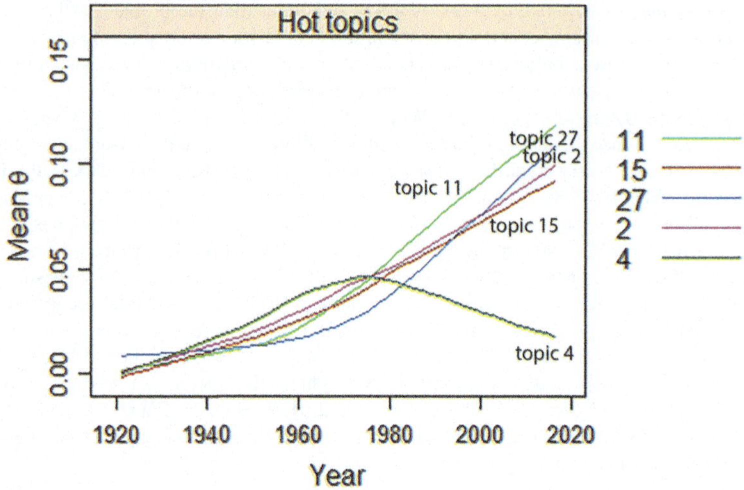

Fig. 3.5 Hot topics in *American Journal of Sociology*

Table 3.6 Top terms in hot topics (decreasing order of probability)

Hot topics				
Topic 11	Topic 15	Topic 27	Topic 2	Topic 4
women	power	violence	racial	variables
gender	network	protest	black	measures
men	exchange	democratic	white	hypotheses
organizational	networks	law	race	variable
organizations	elites	collective	whites	causal
career	actors	legal	blacks	correlations
job	control	labor	inequality	participation
employment	relations	local	segregation	school
jobs	organizational	movement	ethnic	variation
employees	mobilization	critical	education	control
discrimination	organizations	events	status	models
labor market	members	resistance	minority	indicators
female	elite	institutional	neighborhoods	family
workers	corporate	production	negroes	measurement
employers	resources	social movements	negro	social capital
wage	firms	international	discrimination	regression
occupations	business	organizations	neighborhood	hypothesis
earnings	positions	public	americans	delinquent behavior
sex	interests	democracy	schooling	boys
workplace	management	authoritarian	areas	marital dissolution

Gender and discrimination in different social contexts, including jobs and career opportunities, is the topic showing the most significant growth (topic 11). The 2009 abstract for the paper "Cumulative Gender Disadvantage in Contract Employment" has the highest level of probability for this topic (99.2%):

> "Women's wages do not grow with experience or tenure as much as men's do. Many accounts of this cumulative gender disadvantage attribute it to women's underinvestment in firm-specific skills. Yet if that were true, this disadvantage would not exist where firm-specific skills are not rewarded by the labor market. This article investigates this argument in the context of contract employment, where demand for firm specificity is minimal. Contrary to expectations, men still receive higher rewards than women over time. Drawing on quantitative evidence and qualitative fieldwork using job histories of high-skill contractors affiliated with a staffing firm, the author finds support for two sources of women's disadvantage: lower rates of movement across clients on the supply side and unmeasured demand-side factors by which similar levels of tenure and client transitions accrue lower rewards to women. Implications for research on gender stratification and career advancement in non-formalized labor markets are discussed" (Fernandez-Mateo 2009, p. 871).

"Women" as a topic was close to zero in the AJS abstracts during the 1950s. Here, it should be pointed out that "Sociologists for Women in Society" was founded only in 1970, while many journals, including the ASA, long resisted women sociologists' efforts to conquer a space of their own. Likewise, gender studies fought a lengthy battle for access to the *AJS*, as the topic appeared only in the mid-to-late 1980s. Today, gender issues are at the core of change but their increasing presence as a hot topic should not be over-estimated. In terms of recognition, women sociologists still have a long way to go. In her second centennial essay on academic reputation, AJS's editor Elisabeth S. Clemens discussed the publication choices made by AJS and the American Sociological Review (ASR) in a 1-year period (1987–1988), noting that AJS, "often perceived as the more "literary journal" (…) appears more open than ASR with regard to content and style of sociology. In effect AJS serves as a "big tent" for diverse methodologies" (Clemens et al. 1995, p. 467). She drew attention to one of the characteristics which continues to put women's careers at the margins of mainstream sociology, where the quantitative approach is the paramount evidentiary base: "To note only the most striking results, women were over eight times as likely to be first authors of qualitative pieces in AJS" (ivi., p. 471). Even if the gender issue is progressing, women remain "outsiders" because they are still marginalized in social theory and quantitative methods. This lack of recognition is also linked to the discipline's increasing specialization. What you gain in a subfield arena you lose in terms of theoretical visibility. In his paper in AJS, Collins concludes "that we need a far more political theory of the basis of sexual inequality, of gender ideologies, and the ways these mesh with the economic structure, both domestic and public." (Collins 1986, p. 1554).

In a similar way, articles which discuss racial differences (topic 2) have increased since the 1960s.

> "Persistent racial residential segregation is often seen as the result of preferences: whites prefer to live with whites while blacks wish to live near many other blacks. Are these neighborhood preferences color-blind or race conscious? Does neighborhood racial composition have a net influence upon preferences, or is race a proxy for social class? This article tests

> the racial proxy hypothesis using an innovative experiment that isolates the net effects of race and social class (…). The authors find that net of social class, the race of a neighborhood's residents significantly influenced how it was rated. Whites said the all-white neighborhoods were most desirable. The independent effect of racial composition was smaller among blacks, who identified the racially mixed neighborhood as most desirable. Further, whites who held negative stereotypes about African-Americans and the neighborhoods where they live were significantly influenced by neighborhood racial composition. None of the proposed social psychological factors conditioned African-Americans' sensitivity to neighborhood racial composition" (Krysan et al. 2009, p. 527).

An interesting topic concerns several social issues that emerged during the period considered, and in particular the relationship between delinquency and certain social factors (topic 4). This is a tradition of study for American sociologists of deviance and crime which proposes a critique of the theories and methods that argues that such social dimensions as social capital and family conditions have a greater influence on delinquency than biological characteristics.

> "The hypothesis that IQ is an important variable in explaining delinquent behavior among juveniles is examined theoretically and empirically. From a structuralist perspective, delinquent behavior is a consequence of social institutional practices, rather than of individual characteristics. The correlation of IQ with delinquency is not because IQ exerts any casual influence on delinquent behavior but because, in certain institutional settings (the schools), it may be selected by the institution as a criterion for differential treatment. Changes in institutional practices produce a change in the relationship between IQ and delinquency. Empirically, the variables in the structuralist model developed by the Office of Youth Development explain over 20% of the variance in serious and non-serious delinquency. The variables used in the IQ-delinquency hypothesis, a model based on individual characteristics instead of on institutional practices, explain less than 5% of the variation in serious and non-serious delinquent behavior. The conclusion is that the IQ-delinquency hypothesis contributes nothing to existing delinquency theory" (Menard and Morse 1984, p. 1347).

The study of power in its various facets (social, political, economic) and the use of strategic networks to gain influential power, as well as corporate governance practices for allocating power and control, is another area that has attracted increasing interest (topic 15).

> "The issue addressed is, what does a social group need to gain political power? Empirical, historical analysis is utilized to explore the relative saliency of three determinants of political power to explain variation in power as defined conceptually and operationally independent from its determinants. The determinants are magnitude, the extent to which the group is tied into economic, political, and social networks, and the degree of mobilization. (…) However, the effect of network relations was notably weak, contrary to predictions of the power-elite perspective. The conclusion stresses the contingent nature of the determinants of political power" (Roy 1981, p. 1287).

> "Corporate governance describes practices that allocate power and control within public corporations, especially between shareholders, the board of directors, and managers. Shareholder value norms have replaced earlier managerialist governance models. (…) These findings have implications for theory and research on collective action in corporate governance" (Benton 2016, p. 661).

Lastly, studies of collective protest and social process among mobilized groups (topic 27) have grown since the 1960s and 1970s, when these phenomenon entered the American sociological agenda.

"Social movements occupy a shared ideational and resource space, which is often referred to as the social movement sector. This article contributes to the understanding of the relational dynamics of the social movement sector by demonstrating how ideational linkages are formed through protest events. Using a data set of protest events occurring in the United States from 1960 to 1995, the authors model the mechanisms shaping why certain movement issues (e.g., women's and peace or environmental and gay rights) appear together at protest events. They argue that both cultural similarity and status differences between two social movement issues are the underlying mechanisms that shape joint protest and the resultant ideational linkages between issues. Finally, they show that the linking of issues at protest events results in changes in the prominence of a given issue in the social movement sector" (Jung et al. 2014, p. 187).

The period of the hot topics overlaps with the expansion of the departments of sociology in American universities, the assertiveness of sociologists as professional figures and the arrival of Functionalism theory, which plays an important role in integrating the discipline. As Connell points out in AJS (Connell 1997, p. 1545), the teaching of the canon—the "classical theory" of which Parsons and Mills were the most prominent architects—in American graduate education consolidated the ideology of professionalism in sociology that the empiricists of the 1920s struggled to establish. It is also a period where the pre-eminence of statistical analysis and functionalism were criticized, encouraged by the emergence of social protest movements in society at large during the 1970s and 1980s. Howard S. Becker's work "Boys in White" (Becker et al. 1961) and Fisher's criticism of the absence of research testing theory in his 1995 article "Towards a Subcultural Theory of Urbanism" (Fischer 1975), attest to the vivacity of the internal debate, but also underscore the risk of fragmentation of the discipline, at the expense of a more universalized conceptual framework. As hot topic 15 shows, *AJS* gave space to the boom of economic sociology in the 1990s, and network analysis gradually found its way into the journal's pages not only as an empirical means of analysing power, but also as an attempt to build a theoretical framework for institutional social interactions. These new issues run parallel to a new interest in the struggles of civil society, with renewed scrutiny of race relations and minorities, gender conditions in particular. It appears that sociology's destiny continues to be linked to the diversity and heterogeneity which provide its complexity and richness, but also leave the discipline with a weak core whose boundaries are by no means clear. Strengthening this core, and mapping these boundaries, is a challenge that continues today.

3.5 Conclusion: Sociologically Speaking, Where Do We Come from and Where Are We Going?

Applying topic detection to articles published in mainstream journals which mirror the sociological debate in a specific historical moment is a new way of reading the history of a discipline. We analysed the trends shown by topics emerging from a text corpus to shed light on some of the shifting directions taken by sociological inquiry in the period considered. AJS has come a long way from its early period as the voice of the new discipline of sociology, when it sought to differentiate itself from theology in the first place and then from other social sciences such as social psychology, and later from economics and social work. Though the days of the Chicago School are now long past, Craig Calhoun in his introduction to the history of American sociology reminds us that "Sociology emerged as a discipline more at the University of Chicago than anywhere else. There it combined, also perhaps more than anywhere else, philosophy and the history of social thought with a close relationship to social reform, Christian socialism, and ideals of ethically informed action in the city. It became less politically engaged and activist as it became more disciplinary and professional—and as a new generation led by Robert Park and Ernest Burgess distinguished themselves from their founding predecessors" (Calhoun 2007, p. 20). Even if the boundaries between disciplines were not always clear until the canonical period of the 1960s, these porous frontiers left behind two precious traditions: first, the closeness to anthropological methods and ethnological studies which for years informed studies of American communities and racial differences (see hot topic 2), symbolic interactionism, and later of ethnomethodology; second, the association of scientific work with quantification throughout the discipline's history. This is a mainstream concern that sustained all research for many years and enabled American sociology to gain credibility with policy makers, obtaining significant government funding and an influence on social policies (e.g., for deviance, education, and the welfare system). Large-scale surveys were at their apex during the Golden Age under the direction of Merton and Lazarsfeld's Bureau of Applied Research at Columbia, while at Harvard advances in multivariate statistics and path analysis under Duncan and Peter Blau's influence paved the way to new methodological tools.

While sociology continued to expand with the emergence of new sub-disciplines such as gender studies and the sociology of racism in the 1970s and 1990s, a growing debate appeared around the need for more reflexivity on the discipline's history and theory, along with a critical sociology which began (from Gouldner to Cox or Collins) to question the marginality of black issues or the lack of a segmental approach to class, gender, and race.

It is difficult to predict what the content of the special issue for AJS's 150th anniversary might be. Considering the trends in the hot topics discussed here and the continuing methodological controversy, one could imagine that the risks of extreme specialization that the discipline faces today could on the agenda. The journal has retained the open-mindedness of its founders, but seems to struggle to match the glories of the first 25 years or those of the Golden Age. Collins in a 1986 paper in AJS took up this challenge: "What makes a science living and vital is not just that it has some exemplar of past research that subsequent scientists can follow. Its power comes from the fact that it is an intellectual vision (…) so much of the intellectual work of today consists of commentaries on works of the past rather than constructions that are creative in their own right" (Collins 1986, p. 1342–1343). And he concluded: "What we need is to recast our specialties and to expand new specialties of sociologists who work on the cumulation of principles that apply across areas" (ivi., p. 1355).

To a very real extent, AJS is the living witness of American sociology's history—and not only of that history; thus, even if this study and the methodology it employs cannot claim to provide an exhaustive view of the discipline in this specific context, the trends observed here remind us that "the sociologists of the late 19th century (…) had a sense of adventure, a skepticism about authority, and a breadth of interest, which we could still do with" (Connell 1997, p. 1546). This is just the tip of the iceberg: further analyses will shed light on many more aspects that call for deeper reflection.

References

Becker, H., Geer, B., Hughes, E., & Strauss, A. (1961). *Boys in white—Student culture in medical school*. Chicago: University of Chicago Press.

Benton, R. A. (2016). Corporate Governance and Nested Authority: Cohesive network structure, actor-driven mechanisms, and the balance of power in American corporations. *American Journal of Sociology, 122*(3), 661–713.

Blei, D. M., & Lafferty, J. D. (2009). Topic models. In A. Srivastava & M. Sahami (Eds.), *Text mining: Classification, clustering, and applications*. Boca Raton, FL: Chapman & Hall/CRC Press.

Blei, D. M., Ng, A., & Jordan, M. I. (2003). Latent Dirichlet allocation. *Journal of Machine Learning Research, 3*, 993–1022.

Burawoy, M. (1982). Introduction: The resurgence of Marxism in American sociology. *American Journal of Sociology, 88*(Suppl), S1–S30.

Calhoun, G. (2007). *Sociology in America: A history*. Chicago: University of Chicago Press.

Clemens, E. S., Powell, W. W., McIlwaine, K., & Okamoto, D. (1995). Careers in print: Books, journals, and scholarly reputations. *American Journal of Sociology, 101*(2), 433–494.

Collins, R. (1986). Is 1980s sociology in the doldrums? *American Journal of Sociology, 91*(6), 1336–1355.

Connell, R. W. (1997). Why is classical theory classical? *American Journal of Sociology, 102*(6), 1511–1557.

Faris, E. (1937). The social psychology of George Mead. *American Journal of Sociology, 43*(3), 391–403.

Fernandez-Mateo, I. (2009). Cumulative gender disadvantage in contract employment. *American Journal of Sociology, 114*(4), 871–923.

Fischer, C. (1975). Towards a subcultural theory of urbanism. *American Journal of Sociology, 80*(6), 1319–1341.

Griffiths, T., & Steyvers, M. (2004). Finding scientific topics. *Proceedings of the National Academy of Sciences of the United States of America (PNAS), 101*(Suppl 1), 5228–5235.

Grimmer, G., & Stewart, B. M. (2013). Text as data: The promise and pitfalls of automatic content analysis methods for political texts. *Political Analysis, 21*(3), 267–297.

Grün, B., & Hornik, K. (2011). Topicmodels: An R package for fitting topic models. *Journal of Statistical Software, 40*(13), 1–30.

Jung, W., King, B. G., & Soule, S. A. (2014). Issue bricolage: Explaining the configuration of the social movement sector, 1960–1995. *American Journal of Sociology, 120*(1), 187–225.

Kalekin-Fishman, D., & Denis, A. (2012). *The shape of sociology for the 21st century: Tradition and renewal*. London: SAGE.

Krysan, M., Couper, M. P., Farley, R., & Forman, T. A. (2009). Does race matter in neighborhood preferences? Results from a video experiment. *American Journal of Sociology, 115*(2), 527–559.

Lloyd, A. H. (1922). Ages of leisure. *American Journal of Sociology, 28*(2), 160–178.
Menard, S., & Morse, B. J. (1984). A structuralist critique of the IQ-delinquency hypothesis: Theory and evidence. *American Journal of Sociology, 89*(6), 1347–1378.
Meredith, F. (1922). Sociology applied in the field of health. *American Journal of Sociology, 28*(3), 319–325.
R development core team. (2016). *R: A language and environment for statistical computing [software]*. Vienna: R foundation for statistical computing. Retrieved from http://www.r-project.org.
Ross, F. A., & Truxal, A. G. (1931). Primary and secondary aspects of interstate migrations. *American Journal of Sociology, 37*(3), 435–444.
Roy, W. G. (1981). The vesting of interests and the determinants of political power: Size, network structure, and mobilization of American industries. *American Journal of Sociology, 86*(6), 1287–1310.
Scott, A., & Desfor Edles, A. (2011). *Sociological theory in the contemporary era: Text and readings*. Thousand Oaks: Pine Forge Press.
Shanas, E. (1945). The American Journal of Sociology through fifty years. *American Journal of Sociology, 50*(6), 522–533.
Small, A. W. (1895). The era of sociology. *American Journal of Sociology, 1*(1), 1–15.
Small, A. W. (1905). A decade of sociology. *American Journal of Sociology, 11*(1), 1–10.
Turner, S. (1998). Who's afraid of the history of sociology? *Swiss Journal of Sociology, 24*, 3–10.
Weinberg, E. A. (1974). *The development of sociology in the Soviet Union*. London: Taylor & Francis.
Zhou, X., & Pei, X. (1997). Chinese sociology in a transitional society. *Contemporary Sociology, 26*, 569–572.
Znaniecki, F. (1927). The object matter of sociology. *American Journal of Sociology, 32*(4), 529–584.
Znaniecki, F. (1945). Controversies in doctrine and method. *American Journal of Sociology, 50*, 514–521.

Chapter 4
Histories of Social Psychology in Europe and North America, as Seen from Research Topics in Two Key Journals

Valentina Rizzoli

Contents

4.1 Introduction ... 66
4.1.1 "Short History of Social Psychology" ... 66
4.1.2 Bridge over the Ocean? European and (North) American Social Psychology ... 67
4.1.3 The European Journal of Social Psychology and the Journal of Personality and Social Psychology ... 68
4.2 Method ... 69
4.2.1 The Corpora ... 69
4.2.2 The Lexical Correspondence Analysis ... 72
4.2.3 The Reinert Method ... 73
4.3 European and American Social Psychology as Presented in Their Key Journals ... 73
4.3.1 Most Characteristic Words in EJSP and JPSP over Time ... 73
4.3.2 Topics and Trends in the EJSP ... 76
4.3.3 Topics and Trends in the JPSP ... 79
4.4 Discussion ... 82
4.5 Concluding Remarks ... 84
References ... 85

Abstract The study presented in this chapter compares European and US social psychology through the analysis of papers published by two pivotal journals in the discipline: Journal of Personality and Social Psychology and the European Journal of Social Psychology. Scientific production can be considered a starting point for the study of the history of a discipline as it includes theories, application domains and methods that contribute to delineate its trajectory. All the abstracts (from the first publication to the last one in 2016) of the papers of the two journals were collected. By means of a correspondence analysis, the existence of a latent temporal pattern in keywords' occurrences was explored. Furthermore, in order to detect, retrieve and compare the main topics the journals dealt with over time, an analysis implemented by means of Reinert's method was conducted. The topics evolution

---

V. Rizzoli (✉)
University of Padova, Padova, Italy
e-mail: valentina.rizzoli@phd.unipd.it

A. Tuzzi (ed.), *Tracing the Life Cycle of Ideas in the Humanities and Social Sciences*, Quantitative Methods in the Humanities and Social Sciences,
https://doi.org/10.1007/978-3-319-97064-6_4

along time was thus observed and matched in the two journals. Results showed that the two journals have common trajectories particularly in their inception (among others, studies on aggression and attribution) and more recently (among others, studies on gender by means of implicit measures and culture). However, the distinctive feature that characterises the US social psychology, that is the attention on the individual aspects, and the one that characterises the European one, that is the attention on social aspects, seems to remain.

Keywords History of social psychology · American social psychology · European social psychology · Reinert's method · Scientific literature

4.1 Introduction

4.1.1 "Short History of Social Psychology"

Although it is recognised that "modern social psychology" started after the Second World War both in Europe and North America, social psychology's ancestors and founders can clearly be traced to earlier periods (Farr 1996). Given the long-standing links between two continents' approaches to the discipline (Moscovici and Markova 2006), this chapter will attempt to offer a bridge between the European and North American histories of modern social psychology.

Allport's chapter in the Handbook of Social Psychology (1954, 1968, 1985) has long been considered as an "official history", and stated a "long past" of social psychology, mentioning its European roots, and its "short history", referring to the time in which it has become an experimental science, mainly in North America. Although the importance of Allport's viewpoint is widely acknowledged, the chapter has been criticised for presenting an incomplete and misrepresented picture (Farr 1996; Lubek and Apfelbaum 2000) that serves to reproduce the rhetoric of social psychology as an "American phenomenon" (Allport 1954). In the fifth edition of the Handbook, Ross et al. (2010) introduce the chapter on the history of social psychology by noting that they are describing "a" history, rather than "the" history. I will start this chapter with a similar premise. In considering social psychology as a historical product of institutions, practices and beliefs (Danziger 1995), how it is disseminated (e.g. through handbooks or papers) is fundamental in shaping it. As suggested by the seminal theme of this book, I propose to describe a history of social psychology on two historically linked continents, as reflected in the scientific debates surrounding the theories, fields of application and methods that have appeared over time in major relevant scientific journals, delineating the discipline's trajectory (Trevisani and Tuzzi 2015, 2018). With this intent in mind, the following introduction will present a "short history" of the intertwined dynamics of European and North American social psychology.

4.1.2 Bridge over the Ocean? European and (North) American Social Psychology

After the Second World War, what is known as "modern social psychology" was developing in North America, starting from the work of Lewin (who had immigrated from Berlin to the United States) and his students. This was referred to by Moscovici and Markova (2006) as the "indigenous-American tradition", as opposed to its newer Euro-American counterpart. At that time, in fact, a European tradition had not yet been established, although many scholars were working, mainly independently, on social psychology (cf. Moscovici 1999; also Kruglanski and Stroebe 2012). First in 1963, and then in 1964, John Lanzetta promoted two conferences funded by the U. S. Office of Naval Research to re-unite social psychologists in Europe. It was here that the idea was conceived which led to the European Association of Social Psychology (EASP—initially called the European Association of Experimental Social Psychology), founded in 1966 (see Graumann 1995/1999; Moscovici and Markova 2006). The Association then aimed, as it still does today, to promote a distinctively "European" brand of the discipline to the rest of the world in general and to North America in particular.[1]

As "American[2]" and "European" social psychology cannot be considered as two completely separate and counterpoised entities, however it is necessary to remind what in those labels is implied. Essentially, the distinction between them hinges on the fundamental question on which the discipline was established: the relationship between the social and individual dimensions in determining what people think, feel and do (Contarello and Mazzara 2000, p. 5). "American" social psychology stems mainly from the indigenous-American tradition expressed in Floyd Allport's 1924 textbook, where social psychology is considered part of general psychology, i.e. there is more attention on the "individual". By contrast, "European" social psychology refers to a tradition, promoted by the EASP, which assigns a larger role to social and cultural aspects. In the European view, social psychology is strictly connected to disciplines such as sociology and anthropology, and the attention is more on the "social" (see Palmonari and Emiliani 2014). Nevertheless, it is important not to stop at the labels and to bear in mind that there are also a number of different approaches that were originally proposed by North American scholars but are not individually focused (e.g. socio-constructionism, among others) or American researchers who follow the European tradition, as well as European researchers who follow those American.

Several scholars have offered comparative discussions of American and European social psychology, mainly by reflecting on the distinction outlined above (see Jaspars 1986; Moscovici and Markova 2006). Scherer (1992, 1993), for instance, referred to the European and American social psychology as the *two faces* of the

[1] http://www.easp.eu/about/?

[2] The broad label "American social psychology" is used in reference to North America, the USA in particular.

discipline, stressing their differences and the points they have in common. In an extensive study that involved conducting mail survey with social psychologists, comparing textbooks and updating previous works (see Fisch and Daniel 1982; Jaspars 1986), Scherer pointed out that European researchers are primarily interested in intergroup relations, social identity and social influence in terms of group factors, whereas North American scholars are mainly concerned with individuals and their relationships to social environment. Moreover, the scholar identified self-awareness, interpersonal attraction and personal relations as key developments in North America, and language, as part of social communication, in Europe. Common interests include emotions and affects.

4.1.3 The European Journal of Social Psychology and the Journal of Personality and Social Psychology

This chapter will trace a history of social psychology in two historically linked continents, as reflected in the articles published over time in two major mainstream journals: The European Journal of Social Psychology (EJSP) and the Journal of Personality and Social Psychology (JPSP).

The former is an official publication of the EASP, and, as the voice of the Association, is considered a crucial step in enhancing European social psychology's international visibility.[3] It was originally published by Mouton & Co., and in 1977 was taken over by John Wiley & Sons, its current publisher. The journal arises as an international forum in all areas of social psychology. The JPSP is put out by the American Psychological Association, the largest community of psychology in the United States, and is also important in providing guidelines in Europe. In terms of visibility, the JPSP is considered one of social psychology's most important journals, "a flagship journal in its area" (Tesser 1991, p. 349). The journal was born as a division from the Journal of Abnormal and Social Psychology (Katz 1966), and since the two journals initially shared the editorial board, in the first issues of JPSP it is possible to find some papers accepted in the Journal of Abnormal and Social Psychology (Katz 1965). Since 1980, the JPSP has consisted of three independently edited sections: Attitudes and Social Cognition, Interpersonal Relations and Group Processes, and Personality Processes and Individual Differences.

This chapter outlines a new approach which applies a distant reading perspective (Moretti 2013) to tracing the histories of American and European social psychology starting from their pivotal scientific production. In particular, it will portray and compare the temporal pattern of the main concepts discussed in EJSP and JPSP and the main topics these journals have dealt with over time.

[3] http://www.easp.eu/about/?

4.2 Method

4.2.1 *The Corpora*

All available references, titles and abstracts in the two journals were collected, and the abstracts included in two corpora. Abstracts are an effective vehicle for the main contents of an article, since they should be written in such a way as to summarise all relevant information (the abstract should make the aims, methods, findings and field of application clear to the reader at a first glance). As a text genre, moreover, they show interesting linguistic features (standardisation, compactness, conciseness) that lend themselves to an analysis based on word frequencies in a bag-of-words framework. For both journals, abstracts were available for papers published since the first issue. For the EJSP, the corpus was collected by merging and comparing information available from different online sources and directly from the journal's website. A total of 2,559 items was collected, from the very first in 1971, Volume No. 1, Issue No. 1 to the latest in 2016, No. 46, Issue No. 7. A period of 46 years was thus covered. Also for the JPSP, all abstracts were collected from online sources and were checked on the journal's website. A total of 9,568 items was collected, from 1965, Volume No. 1, Issue No. 1 to 2016, No. 111, Issue No. 6, for a period of 52 years. Items which did not contain an abstract were deleted (e.g. editorials, master heads, errata, acknowledgements). The EJSP corpus consists of 2,195 abstracts, while the JPSP corpus totals 9,536 abstracts (Table 4.1).

Each corpus contains, together with the abstracts, the information of the corresponding volume and issue. As regards EJSP 46 volumes (one per year) and 251 issues are included. Up to volume 17 there are 4 issues per year. Up to the volume 18 they are 6 per year, and up to the volume 38 they are 7. Hence, the number of publications increases over years (see Table 4.2).

Table 4.1 Overview of the corpora

Journal	Time span	Volumes	Issues	Number of abstracts
EJSP	1971–2016	46	251	2,195
JPSP	1965–2016	111	624	9,536

Table 4.2 EJSP corpus: Volumes, issues, number of available abstracts. Size in word tokens (*N*)

Years	Volumes	Issues	Abstracts	*N*
1971–1975	1–5	20	114	18,255
1976–1980	6–10	20	112	18,878
1981–1985	11–15	20	141	21,124
1986–1990	16–20	26	178	24,842
1991–1995	21–25	30	219	35,243
1996–2000	26–30	30	281	42,796
2001–2005	31–35	30	249	39,510
2006–2010	36–40	33	422	63,593
2011–2016	41–46	42	479	76,170

For the JPSP, 111 volumes and 624 issues (12 per year) are included (see Table 4.3). There is more than one volume per year (they are irregular until 1980, then, starting from that year, there are two volumes per year). To facilitate the comparison only the year has been included as variable in the analysis.

To improve the homogeneity of the corpora in the EJSP, which accepts both American English and British English in the papers, we opted for British spelling (e.g. the word *analyzed* was replaced with *analysed*). By contrast, JPSP uses only American English, which was thus maintained. This was possible because the two corpora were analysed separately. Both corpora were normalised by replacing uppercase with lowercase letters using TaLTaC2 software. The EJSP and JPSP corpora consist respectively of 340,411 and 1,408,911 word-tokens (total number of occurrences for each corpus), and 13,110 and 24,770 word-types (number of different forms). The lexicometric measures showed good redundancy of the corpora (Lebart et al. 1998; Tuzzi 2003; Bolasco 2013), as indicated in Tables 4.4 and 4.5. In fact, the type/token ratio (number of distinct forms divided by the total number of occurrences) is 3.9 in the EJSP and 1.8 in the JPSP and hapax legomena (forms that appear only once) account for 35.9 in the EJSP and 32.9 in the JPSP.

Multiword Expressions (MWEs), i.e. meaningful sequences of words, were identified, selected manually and considered as textual units. Only MWEs with

Table 4.3 JPSP corpus: Volumes, issues, number of available abstracts. Size in word tokens (*N*)

Years	Volumes	Issues	Abstracts	*N*
1965–1970	1–16	72	1176	149,105
1971–1975	17–32	60	1119	152,751
1976–1980	33–39	60	872	131,717
1981–1985	40–49	60	1221	188,730
1986–1990	50–59	60	1109	180,825
1991–1995	50–69	60	871	115,630
1996–2000	70–79	60	926	124,804
2001–2005	80–89	60	742	97,098
2006–2010	90–99	60	737	115,052
2011–2016	100–111	72	763	153,199

Table 4.4 Lexicometric measures of EJSP

N—Word-tokens	340,411
V—Word-types	13,110
(V/N)*100—Type/Token ratio	3.9
(V_1/V)*100—Hapax percentage	35.9

Table 4.5 Lexicometric measures of JPSP

N—Word-tokens	1,408,911
V—Word-types	24,770
(V/N)*100—Type/Token ratio	1.8
(V_1/V)*100—Hapax percentage	32.9

Table 4.6 Excerpt of contingency table words × years. Occurrences in EJSP corpus

Words	Occurrences (corpus)	1971	1972	1973	:	1991	1992	:	2014	2015	2016
the	19,127	246	244	295	:	394	334	:	584	593	445
of	14,023	163	198	180	:	300	282	:	417	461	355
study	1,630	6	13	12	:	20	18	:	71	90	74
participants	1,058	0	0	1	:	4	0	:	47	46	35
experiment	727	8	14	15	:	17	15	:	21	12	14
effects	702	6	3	6	:	16	11	:	35	21	26
people	687	0	1	7	:	7	16	:	41	38	39
ingroup	578	2	0	0	:	3	15	:	26	23	16
information	567	0	2	1	:	8	19	:	11	0	4
outgroup	508	4	0	1	:	5	13	:	25	14	11
identification	244	0	0	2	:	2	6	:	9	26	11
judgments	243	8	4	2	:	3	5	:	13	4	6
tested	243	1	2	1	:	5	1	:	10	10	10
cognitive	242	5	3	1	:	6	5	:	5	10	1
social identity	156	1	0	0	:	0	3	:	11	14	15
partner	145	0	3	1	:	0	3	:	7	3	3
discrimination	145	0	1	1	:	1	1	:	0	3	4
conflict	144	1	0	4	:	0	3	:	13	3	6
message	144	0	0	0	:	2	1	:	5	5	0
political	141	1	0	3	:	0	0	:	7	13	17
paradigm	123	0	1	1	:	6	2	:	0	0	4
stereotype	123	0	0	1	:	8	2	:	0	0	2
social psychology	110	0	0	0	:	5	0	:	4	7	0
sample	109	0	1	1	:	0	1	:	8	5	6
causal	109	0	0	0	:	2	1	:	4	5	1
priming	106	0	0	0	:	0	0	:	6	3	2
implicit	106	0	0	0	:	2	0	:	1	11	4
gamble	1	0	0	1	:	0	0	:	0	0	0
prose	1	0	0	1	:	0	0	:	0	0	0

frequencies higher than or equal to 5 for the EJSP corpus and higher than or equal to 10 for the JPSP (since it is significantly larger than the former) were identified. An automatic information retrieval procedure was used to recognise repeated sequences of words (e.g. an adjective followed by a noun, as in *social identity*) that produce an MWE (see Chap. 8). Further MWEs were found by means of two encyclopaedias of social psychology (Manstead et al. 1995; Baumeister and Vohs 2007) and an index of keywords for the retrieved papers available in Scopus. A total of 997 MWEs were identified in the EJSP and 1,385 in the JPSP. An extract of the vocabularies created and used for the analyses is shown in Tables 4.6 and 4.7.

Table 4.7 Excerpt of contingency table words × years. Occurrences in JPSP corpus

Words	Occurrences (corpus)	1965	1966	1967	:	1989	1990	:	2014	2015	2016
the	70,476	1800	1524	1570	:	1501	1590	:	970	863	784
of	56,673	1393	1137	1177	:	1208	1281	:	787	817	647
study	5,980	70	41	35	:	106	141	:	199	213	197
effects	2,968	67	59	41	:	62	69	:	54	50	49
behavior	2,726	30	55	39	:	39	56	:	38	59	22
participants	2,383	0	1	1	:	6	14	:	76	96	50
information	2,317	49	19	50	:	56	71	:	37	23	8
effect	2,276	52	40	50	:	42	44	:	49	62	40
experiment	2,053	49	33	36	:	68	61	:	33	29	18
performance	1,704	75	50	39	:	43	28	:	11	14	18
hypothesis	1,676	51	36	32	:	44	42	:	21	19	18
subjects	1,583	3	0	1	:	67	111	:	1	0	1
social	1,570	27	18	13	:	26	30	:	41	60	24
women	1,546	18	9	12	:	34	44	:	18	22	44
perceived	1,542	13	13	19	:	38	41	:	35	43	26
relationship	1,502	36	23	36	:	24	20	:	59	41	27
conditions	1,501	70	54	50	:	26	24	:	5	8	10
individual differences	552	4	12	3	:	10	11	:	8	30	21
reward	550	19	23	25	:	1	1	:	0	3	4
memory	529	1	2	1	:	22	30	:	0	4	2
approach	528	0	4	8	:	7	4	:	12	21	8
personality traits	284	1	1	1	:	4	2	:	10	19	24
validity	276	5	2	2	:	5	7	:	2	4	2
construct	276	0	0	2	:	3	8	:	5	6	6
consistently	275	5	4	4	:	1	5	:	5	9	5
empathy	270	0	0	0	:	10	5	:	15	3	2
couples	266	0	0	1	:	2	8	:	5	7	8
defenders	1	1	0	0	:	0	0	:	0	0	0
agers	1	1	0	0	:	0	0	:	0	0	0

4.2.2 *The Lexical Correspondence Analysis*

A (lexical) correspondence analysis (CA) was conducted using the SPAD software package to test for a chronological dimension and to provide a general graphic overview of the structure of the association between years and words. The CA is based on the vocabulary that for each word comprises occurrences in the different years, representing the reference time points. It recognises similarities and differences through the lexical profiles, that is, through the frequencies of words over time (see Chap. 1—Annex). This analysis employed a matrix of 5,784 words over 46 years (rows per columns) for the EJSP corpus and 8,349 words over 52 years for the JPSP corpus. As regards the former matrix, only words with frequencies higher than or equal to 5 were considered, while for the latter matrix, only words with frequencies higher than or equal to 10 were inserted. Keywords and MWEs have been labelled inside a column set as supplementary variable. In this way, the main words (including methods, theories and fields of application) characterising each journal over time were displayed.

4.2.3 The Reinert Method

The two corpora were exported from TaLTaC2 after normalisation and the recognition of keywords and MWs. The main topics covered in the journals were then identified by using the Reinert method (Reinert 1986) with the IRaMuTeQ software package (see Chap. 10). In this approach, topics are defined as "lexical worlds", i.e. groups of words referring to a class of meaning (Reinert 1993). The words that make up a topic are identified on the basis of their co-occurrence in the elementary unit of context (ECU), i.e. the basic statistical units contained in the initial unit of context (in this case the entire abstract). The result, obtained with a hierarchical descending classification, is a dendrogram that groups units into classes that mirror a similar lexical context (Sbalchiero and Tuzzi 2016). For the sake of clarity in presenting and comparing results, a medium level of specificity of the topics (more or less grouped) was chosen (around ten for each journal). IRaMuTeQ output was analysed with the R software package to observe the presence of the identified topics over time (see Ratinaud 2014).

4.3 European and American Social Psychology as Presented in Their Key Journals

4.3.1 Most Characteristic Words in EJSP and JPSP over Time

The lexical correspondence analysis indicated that there is a chronological dimension in both the EJSP and JPSP corpora (Figs. 4.1 and 4.2). The main contents of the abstracts of articles published in the journals can be described following the timeline of the reference periods across the four quadrants.

For the EJSP, the first two axes (out of 45) represent 9.57% of explained inertia (6.10% for the first-horizontal axis and 3.47% the second-vertical axis). The top-left quadrant (Fig. 4.3) groups together the first 20 years of the journal's life (1971–1991). These years are characterised by forms that refer to studies on *aggression, dissonance, attribution* and *cooperation/competition*, as well as words referring to methods, such as *factorial design* and *experimental.*

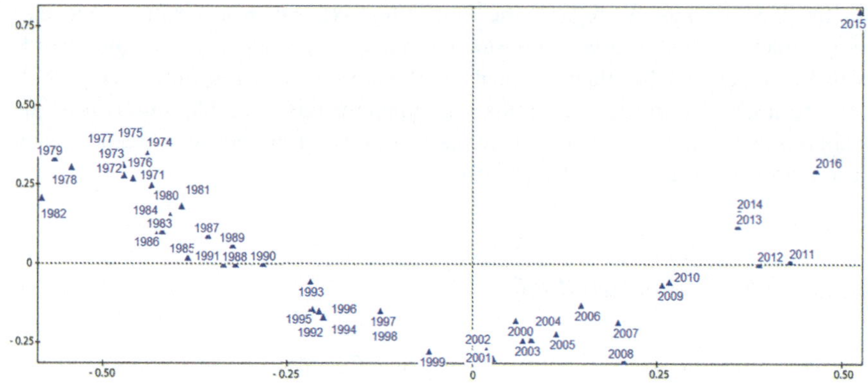

Fig. 4.1 First factorial plane of the EJSP correspondence analysis. Projection of years

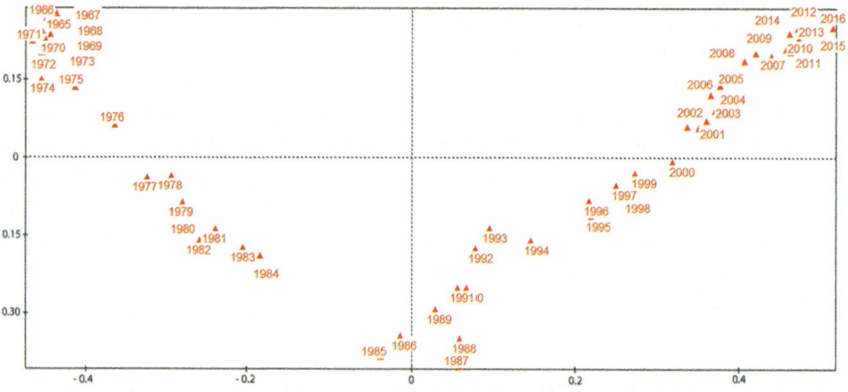

Fig. 4.2 First factorial plane of the JPSP correspondence analysis. Projection of years

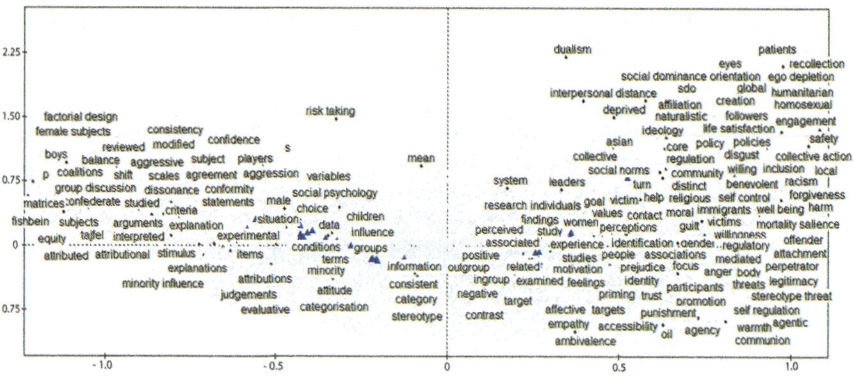

Fig. 4.3 First factorial plane of the EJSP correspondence analysis. Projection of the 6% of words with the highest contribution

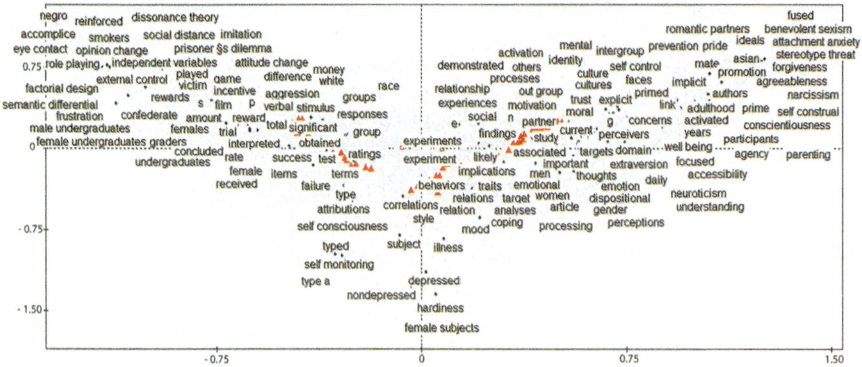

Fig. 4.4 First factorial plane of the JPSP correspondence analysis. Projection of the 5% of words with the highest contribution

> The influence of two different reinforcers of **aggression** was investigated: Frustration (intrinsic primary reinforcement) and instrumental value of **aggression** (extrinsic primary reinforcement). In the first part of the **experiment** frustration was manipulated on two levels by having the stooge interfere very often or seldom in the ,building a village' task of the **subject**.
>
> Quotation from abstract, 1973, 3(2)

In the subsequent years (mainly the mid-1990s), some of the hallmarks of European production (e.g. *minority influence*) appear. Starting from the immediately subsequent years (late 1990s), we find words that refer to the cognitive turn (e.g. *stereotype, categorisation*), together with *motivations* and intergroup processes.

> Two field studies investigated whether as predicted by **self categorisation theory** (Turner 1987), the relationship between comparative fit of an **ingroup outgroup categorisation** and group phenomena is mediated by depersonalisation of self perception, and moderated by category accessibility.
>
> Quotation from abstract, 2006, 36(1)

The last period (since 2010) seems be characterised by the presence of various social issues, such as *migration, religion* and *gender*. Moreover, words such as *mediated* that refer to the role of variables mirror an orientation towards experimentation.

> This article analyses the influence of accent on discrimination against **immigrants** [...] we replicated this effect and found that the influence of accent on discrimination is **mediated** by the perceived quality of the accent.
>
> Quotation from abstract, 2016, 46(5)

For the JPSP, the first plane (first two axes out of 51) represents 19.65% of explained inertia (13.94% for the first-horizontal axis and 5.71% for the second-vertical axis). In the first years considered (1965–late 1970s, Fig. 4.4), words such as *aggression, frustration, attitude,* others referring to methods (e.g. *factorial design, s—subjects, independent variables)* and others that refer to behaviourism (e.g. *reinforcement*) are typical.

> In 3 **experiments Ss**, **male college students**, were either angered or treated in a neutral fashion by a person who had been labeled either as a college boxer or a speech major, and they were then shown a short film. […] The findings in **Exp**. III confirmed the results obtained in earlier investigations by showing that the angered **Ss's** inhibitions against **aggression** varied with the apparent justification for the observed **aggression**.
> Quotation from abstract, 1965, 2(3)

In the subsequent years (1980s), words linked to personality and personality tests are characteristic (e.g. *scale, type a*). In the following years (the 1990s), personality continues to enjoy primacy, with words such as *depression, coping* and *trait*. Words referring to the cognitive turn (e.g. *memory*) and *emotion* are also characteristic.

> This study tested an integrated interpersonal theory of **depression** […] **Depressed** targets reported engaging in more negative feedback seeking than **nondepressed** targets, and tended to report seeking more reassurance than **nondepressed** targets.
> Quotation from abstract, 1995, 69(4)

The last years considered are still characterised by personality (e.g. *big five, conscientiousness*) and words referring to relationships (e.g. *partner, relationship*), as well as by words more related to social aspects such as *culture, groups* and *identity*.

> Accurately perceiving whether interaction **partners** feel understood is important for developing **intimate relationships** and maintaining smooth interpersonal exchanges. During **interracial interactions**, when are Whites and **racial minorities** likely to accurately perceive how understood **cross race partners** feel?
> Quotation from abstract, 2015, 108(1)

4.3.2 Topics and Trends in the EJSP

Nine topics were identified in the EJSP corpus (Table 4.8) using Reinert's method. Topics are composed by groups of words, that are the most relevant of the ECUs (the abstracts) classified in that topic. All together, the topics account for 79.74% of the abstracts (1,748 abstracts out of 2,195). Since the aim was to classify the journals' abstracts into a limited number of clear-cut and easily interpretable clusters, it was decided to consider each abstract as a unit of context (as a whole), which is the starting point of all analyses carried out to identify topics by means of Reinert's method. In this way, it was possible to obtain a clear output even if not all the abstracts can be assigned to a cluster (about 20% for the EJSP corpus and 24% for JPSP remained out). For the sake of clarity and comparability, as mentioned above, it was decided to limit the number of clusters to around ten. To choose the precise number, several trials were carried out with various number of clusters, after which the number closest to ten that provided a good explanatory power was chosen. A brief interpretation of the identified topics will be provided below. To do so, it was necessary to reconnect the representative words in the clusters to their abstracts.

Table 4.8 Dendrogram of EJSP topics

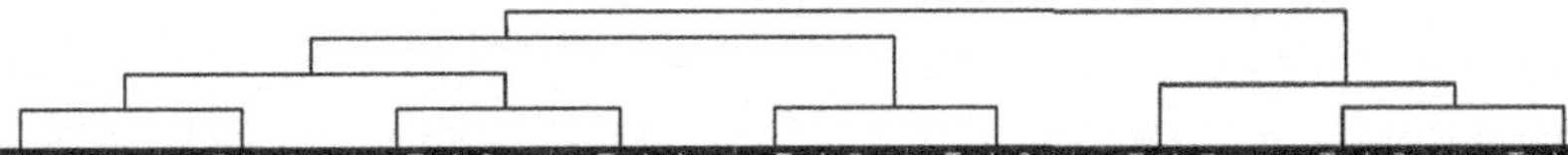

Topic 1	Topic 2	Topic 3	Topic 4	Topic 5	Topic 6	Topic 7	Topic 8	Topic 9
9.7%	13.6%	12.5%	9.1%	12.3%	11.9%	11%	9%	10.9%
attribution attribution_theory causal_attribution causality internal covariation information Kelley method statistical conversational validity model error heuristic representativeness primacy	stereotype stereotype_ content stereotype_ change stereotypic categorisation category categorical memory social_judgment social_ perception judgment evaluatively recall recognition memory impression impression_ formation person_ perception	prime implicit association_test implicit explicit_attitudes IAT stereotype threat man women math performance self_evaluation objectification benevolent_ sexism participants female_ participants photo	mood mood_effect positive_mood neutral_mood happy sad regulatory regulatory_ focus prevention promotion goal diet unhealthy health_ behaviours regulatory_fit eat persuasion persuasive_ message	subject ss factorial_ design analysis_of_ variance dependent_ variable aggression aggressive shock violence frustration group_ polarisation choice dilemma dissonance cognitive_ conflict cognitive_ dissonance	cooperation cooperative competition competitive decision allocate fairness bargain game social dilemma reward negotiation equity prisoner's_ dilemma maximise decision_ making prosocial minimal_ group group_ member intergroup_ competition	political ideology culture western conservativism conservative liberal universalism vote ideological religion religious religious_identity west authoritarianism democracy attitude political_parties Schwartz European New Zealand Germany Triandis right_wing_ authoritarianism system_ justification	ingroup social_identity outgroup self_ categorisation group group_status Turner identity outgroup_ homogeneity intergroup_ identification intergroup_ context ingroup identification ingroup_ favouritism ingroup_ identification	immigrant intergroup contact prejudice contact intergroup_ attitudes national_ identification ethnic_ minorities perceived_ discrimination racial host_society Moroccans Majority_group Minority_group Social_ dominance_ orientation Intergroup_ anxiety

For each topic, some of the words shown to be most characteristic by the chi-square test are listed. All words listed have a p value <0.0001

1. *Attribution*

   The first topic groups together words that mainly refer to attribution theories. For example, references to causal attribution (e.g. *causal, causal attribution, internal*) and the Kelley attribution model (e.g. *Kelley, covariation information*) are representative of this cluster. Words referring to methodological aspects of these studies (e.g. *methodological, statistical, validity, conversational*) also appear. In addition, words referring to heuristics (e.g. *heuristic, representativeness*) are present. It can be seen that this topic was mainly present in the 1970s and 1980s, and, with an isolated peak (containing mainly words referring to heuristics), in 1997 (Fig. 4.5).

2. *Stereotyping and impression formation*

   The second topic is characterised by words that refer to stereotyping (e.g. *stereotype, stereotype content, stereotype change*) and the associated process of social categorisation (e.g. *category, categorical*). Memory, social judgment and social perception (e.g. *judgment, judge, evaluatively, recall, recognition memory*) appear as words defining this topic. In a related area, words referring to impression formation (e.g. *impression, impression formations, person perception*) are also typical. In general, these processes clearly refer to the social cognition perspective (see Fiske and Taylor 1982). This topic was chiefly characteristic of the 1990s (Fig. 4.5).

3. *Implicit measures and gender*

   In the third topic, words relating to implicit measures (e.g. *prime, implicit association test, implicit*) are characteristic. Other words in the same topic refer to where these measures are applied, viz., mainly gender as stereotype (e.g. *man,*

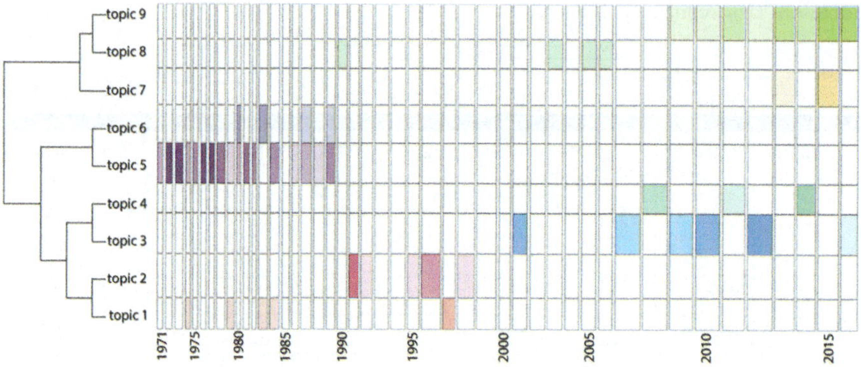

Fig. 4.5 Over-representation of EJSP topics per years. The height of the lines of each class is proportional to the dimension of the class in terms of the number of abstracts it contains. The width of the cells is proportional to the frequency of the ECUs (abstracts) in a given year. The tone of the colour is proportional to the strength of the association between class and year

woman, stereotype threat, self-evaluation, math performance) and to processes related to gender (e.g. *objectification, benevolent sexism*). Moreover, words that refer to the studies setting (*participant, female participants, photo*) are present. This topic gained relevance at the beginning of the 2000s and became prominent in the last decade considered (Fig. 4.5).

4. *Mood and Regulatory Focus Theory*

   The fourth topic contains words related to mood (e.g. *mood effect, positive mood, neutral mood, happy, sad*). Regulatory focus theory (*regulatory focus, goal, promotion, prevention*) and its applications also have a dominant presence. For example, words such as *diet, health behaviours, eat, persuasion* and *consumer* are typical. This topic is prevalent in 2008, 2011, and 2014, i.e. the last decade considered (Fig. 4.5).

5. *Factorial design (aggression, group polarisation, dissonance)*

   The fifth topic is characterised by words referring to factorial design (e.g. *subject, factorial design, analysis of variance, dependent variable*) and different fields where it is used. These include aggression (e.g. *aggressive, violence, frustration*), group polarisation (e.g. *choice dilemma, group polarisation*) and dissonance (e.g. *cognitive conflict, cognitive dissonance*). Studies relating to this topic were dominant from 1971 to 1989 (Fig. 4.5).

6. *Cooperation/competition and game theories*

   The sixth topic contains words referring to studies on cooperation and competition (e.g. *cooperative, competitive*) and, more generally, game theories. In fact, words such as *game, negotiation, decision, allocate, distribution, equity* and *reward* that refer to the dynamics involved in these studies are characteristic, as well as *prisoner's dilemma game*. The processes involved are also present (e.g. *minimal group, social dilemma*). This topic was particularly noticeable in the 1980s (Fig. 4.5).

7. *Politics and (cross) culture*
   The seventh topic mainly contains words that refer to politics and culture. In the former area, they include, for example, *political, ideology, vote, political parties, conservative, liberal.* Words relating to culture are *culturally, cultural value, cultural differences, west, European,* and names of countries. References to religion (e.g. *religious, religious identity*) are also typical. Studies on culture are mainly cross-cultural, as is also confirmed by the presence of the name of the seminal author: *Triandis.* Moreover, those studies investigate how certain characteristics measured by scales (e.g. *right wing authoritarianism, social dominance orientation, system justification*) are related to certain contexts. This topic is mainly present in the last years considered (2013, 2015; Fig. 4.5).
8. *Social identity theory and ingroup/outgroup processes*
   The eighth topic groups together words concerning social identity theory and ingroup/outgroup processes. In fact, together with the name of the theory itself, we find words referring to the processes it involves, including *self-categorisation, ingroup favouritism, ingroup identification, social identification,* as well as words such as *ingroup, outgroup, member* and *status.* This topic was particularly noticeable in 1990 and the mid-2000s (Fig. 4.5).
9. *Intergroup contact and applied concerns*
   The ninth topic is related to the eighth, but the words it contains mainly refer to intergroup contact and applied concerns, including migration (e.g. *immigrant, ethnic, host society, Moroccans*). Related processes are also seen, such as *prejudice, intergroup anxiety* and *perceived discrimination.* This topic is predominant from 2009 until 2016 (Fig. 4.5).

4.3.3 Topics and Trends in the JPSP

Eleven topics were identified in the JPSP corpus using Reinert's method (Table 4.9). As explained above, topics contain the most relevant words of the abstracts classified inside them. All together, the topics account for 76.08% of the abstracts (7,255 abstracts out of 9,536). Topics were labelled using the procedure described above. A brief interpretation is provided below.

1. *Consensus and attribution*
   The first topic contains words concerning studies on consensus (e.g. *false consensus, consensus bias, consensus information*) and attribution (e.g. *attribution theory, external attributions, dispositional, situational*). As regards the latter, characteristic words include those relating to causal attribution (e.g. *causal*), the Kelley attribution model (e.g. *Kelley, covariation*). In general, words referring to bias and heuristics in attribution and judgment processes such as representativeness heuristic and base rate fallacy (e.g. *base rates, Kahneman, Tversky*) and correspondence bias are characteristic. This topic is not clearly prevalent in particular grouped years, though it appears at the end of the 1970s, the end of the 1980s, and at another time in 2004 (Fig. 4.6).

Table 4.9 Dendrogram of JPSP topics

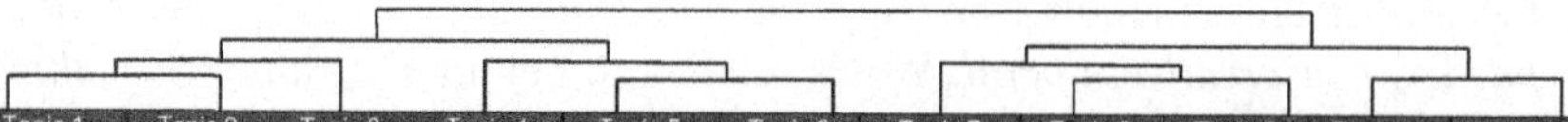

Topic 1 7.6%	Topic 2 9.9%	Topic 3 10%	Topic 4 11.3%	Topic 5 11.1%	Topic 6 10.4%	Topic 7 6.1%	Topic 8 8.8%	Topic 9 7.5%	Topic 10 10.5%	Topic 11 6.8%
inference attribution information base_rate consensus false_ consensus consensus_ bias consensus_ information attribution_ theory external_ attribution dispositional situational causal_ attribution causal Kelley covariation Kahneman Tversky Correspondenc e _bias	memory stereotype category encode information_ retrieval recall recognition_ memory schema inconsistent social_ stereotype stereotype maintenance categorisatio n social category impression formation social information ingrouop outgroup group membership group variability implicit implicit attitudes	self self_ evaluation emotion happiness guilt empathy angry disgust sadness jealousy motivation motive goal motivate feel social_ comparison social_ exclusion empathy- altruism_ hypothesis counterfactu al _thinking	aggression aggressive violence frustration shock arousal skin_ conductance heart_rate electric_ shock electroderma l facial_ expression Ekman emotional_ experience verbal_ response exp female_ undergraduate male_ undergraduate videotape instigation noise	attitude_ change attitude attitude_ statements cognitive_ dissonance dissonance_ theory dissonance_ effect Festinger decision_ making risky_shift group_ discussion choice dilemma persuasive_ message persuasive_ communication persuasive_ arguments source_ credibility exposure_ frequency	ss condition experimenta l reinforcement verbal_ conditioning social_ reinforcement conditioning procedure task reward perform assign reinforcer factorial_ design task_ performance	culture cultural united_states individualism political western collectivism cultural_ differences japan china Spain Mexico east conservativism west liberal conservative social_ dominance authoritarianis m social_distance social_group social_identity intergroup_ relations ingroup group	measure factor analysis psychometric correlation dimension subscale matrix reliability test-retest internal_ consistency discriminant _validity predictive_ validity convergent_ validity incremental _ validity multitrait multimehod self_report subscale personality_ measureme nt self-report_ measures love_scale	masculinity femininity sex_role twin genetic sex Bem_sex_rol e _inventory dizygotic monozygotic big_five personal_ attribute_ questionnaire self_concept_ scale q_set_ inventory gender	well-being stress depression mood_ disturbance health_ problem trauma cope illness depressive_ symptoms stressors physical_ health coping_ strategies social_ support life_events psychologic al_well- being resilience	attachment attachment _security attachment _avoidance Bowlby insecure avoidant couple partner marital wife relationship satisfaction intimate romantic_ relationships dating relationship relationship_ quality ambivalent_ attachment_ style

For each topic, some of the words shown to be most characteristic by the chi-square test are listed. All words listed have a p value <0.0001

2. *Memory, stereotype and categorization*

   The words in the second topic relate mainly to memory, stereotypes and the process of categorisation. In fact, the topic contains words such as *memory, encode, retrieval, recall, recognition memory, store* that relate to memory; *schema, inconsistent, social stereotype, stereotype maintenance*, relating to stereotype and, as regards categorisation, *category, social categorization*, and so on. Here, we find words associated with processes, including *impression formation, social information processes*. This topic also encompasses studies on groups (e.g. *ingroup, outgroup, group membership, group variability*), as well as studies on the implicit (e.g. *implicit, implicit attitudes*). This topic can be seen in particular from the mid-1980s to the mid-2000s (Fig. 4.6).
3. *Self, emotion and motivation*

   The third topic contains words relating to the self (e.g. *self, self-evaluation*), emotion (e.g. *happiness, guilt, empathy, angry, disgust, sadness, jealousy*) and motivation (e.g. *motive, goal, motivate*) and such connected processes as *social comparison, social exclusion, empathy-altruism hypothesis, counterfactual thinking*. This topic is predominant in the last decade considered (Fig. 4.6).
4. *Aggression, arousal and physical measurement*

   The fourth topic is related to studies on aggression (e.g. *aggressive, violence, frustration*) that are mainly conducted by employing physical measurements (e.g. *skin conductance, heart rate, electric shock, electrodermal*) and by monitoring reactions (*facial expression, Ekman, emotional experience, verbal response*). Various words relating to the methods used to conduct these studies,

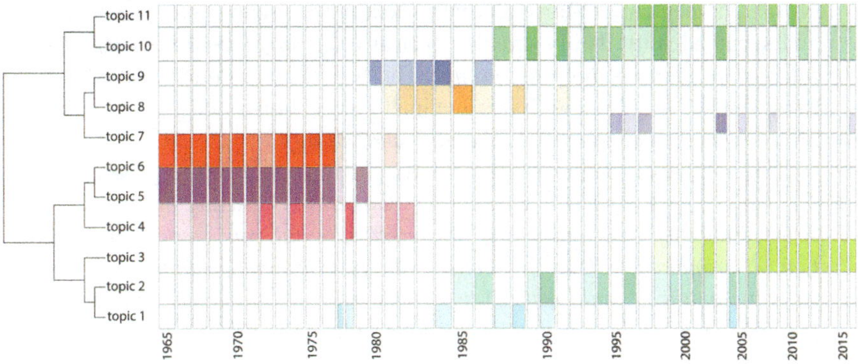

Fig. 4.6 Over-representation of JPSP topics per years. The height of the lines of each class is proportional to the dimension of the class in terms of the number of abstracts it contains. The width of the cells is proportional to the frequency of the ECUs (abstracts) in a given year. The tone of the colour is proportional to the strength of the association between class and year

which are mainly experiments, are also found (e.g. *exp, female/male undergraduate, videotape, instigation, noise*). This topic was particularly prominent in the journal's first years and continued until the early 1980s (Fig. 4.6).

5. *Attitude change, dissonance and decision making*

   The fifth topic mainly contains words that refer to studies on attitude change (e.g. *attitude, attitude statements*), cognitive dissonance (e.g. *dissonance theory, dissonance effect, Festinger*) and decision making (e.g. *risky shift, group discussion, choice dilemma*). Moreover, words linking these processes to communication and persuasion are also characteristic (e.g. *persuasive, message, persuasive communication, persuasive arguments, source credibility, exposure frequency*). This topic was predominant in the first decade of the journal's life (Fig. 4.6).

6. *Experimental condition and procedures*

   The typical words in the sixth topic mainly refer to methods that involve experimental procedures and betray the influence of behaviourism. In fact, words such as *ss* (subjects), *condition, experimental, reinforcement, verbal conditioning, social reinforcement, conditioning procedure*, and so on characterise this topic, which, like the previous one, was in vogue in the first 10 years of the journal's life (Fig. 4.6).

7. *(Cross)culture and politics*

   The seventh topic contains words relating to culture and politics. As regards culture, the characteristic words are more related to cross-cultural studies, as psychological aspects (e.g. *individualism collectivism*) are compared among different cultures (mainly West-East). Words referring to various nationalities are typical (e.g. *United States, Japan, China, Spain, Mexico*). As regards politics, the comparison is mainly between conservative and liberal ideologies. Words such as *political, ideology, liberal* and *conservative*, together with such related processes as *social dominance, authoritarianism, social distance* and so on, are characteristic. Moreover, words relating to groups, intergroup relations and social identity theory are also typical (e.g. *social group, social identity, intergroup relations, ingroup, group*). This topic is intermittently predominant in the mid-1990s, the mid-2000s and in 2016 (Fig. 4.6).

8. *Measurements and construct reliability and validity*

   The eighth topic includes words referring to measurements (e.g. *measure, factor analysis, psychometric, correlation, dimension, subscale, matrix*) and construct reliability (e.g. *reliability, test-retest, internal consistency*) and validity (e.g. *discriminant validity, predictive validity, convergent validity, incremental validity, multitrait multimethod*) associated with questionnaire and self-report measures. This topic was particularly noticeable in the 1980s (Fig. 4.6).

9. *Individual differences*

   Words in topic nine refer mainly to individual differences. For example, characteristics (mainly gender-related, such as *masculinity* and *femininity*, or personality attributes and traits) are studied through twin studies (e.g. *twin pairs, genetic, dizygotic*) and such scales and personality inventories as *Bem sex role inventory, big five, personal attribute questionnaire, self-concept scale, q set* and so on. This topic is linked to the previous one (given the proximity indicated by the descendent hierarchical classification) and, like it, was mainly present in the 1980s (Fig. 4.6).

10. *Well-being and coping*

    The tenth topic contains words relating to well-being and how to cope with negative feelings and situations: *stress, depression, mood disturbance, health problem, trauma*. In fact, words such as *coping strategies, health, social support, life events, psychological well-being, resilience*, and so on are characteristic. This topic gained ground in the late 1980s, and continued to be prominent until the last years considered, though there was some discontinuity, particularly in the 2000s (Fig. 4.6).

11. *Attachment and close relationships*

    The 11th topic includes words referring to attachment (e.g. *attachment security, attachment avoidance, Bowlby, insecure, avoidant*) and close relationships (*couple* and *romantic*). In connection with relationships, for example, we find *couple, partner, marital, wife, relationship satisfaction, intimate, romantic relationships, dating relationship*, and so on. Starting from the late 1990s, this topic has been particularly prominent in the last 20 years considered (Fig. 4.6).

4.4 Discussion

This study first provided an overview of the main concepts that characterised the EJSP and the JPSP over time, and then proceeded to analyse how the topics discussed in the journals have developed through the years. Both stages were useful in order to describe, compare and contrast the EJSP and the JPSP, and also to gain a better understanding of the histories of social psychology, as the journals' historical and contextual relevance makes them pivotal publications in the field.

As can be seen from the results of the correspondence analyses, certain words common to both journals in the first period considered contribute to forming the axes, referring, for example, to dissonance theory and studies on aggression. The main distinction that emerges concerns JPSP's strong focus on personality (mainly starting from the mid-1980s), on the one hand, and work on the psychology of groups by Tajfel and Moscovici in the EJSP, on the other hand. This is particularly true in the 1980s and 1990s. The most recent publications in the two journals show both distinctive and common features. Cognitive processes (e.g. *stereotype* or *bias*) seem to be common features, as well as words that refer to correlational studies and the experimental method employed in the research design (e.g. *experiment*, *factorial design*). Distinctive features include certain issues that principally relate to personality and close relationships for JPSP, and a major emphasis on various social issues (e.g. *environment*, *migration*) for the EJSP.

The results of the analysis conducted with Reinert's method shed light on the main topics dealt with in EJSP and JPSP and their representation throughout the journals' lives. The two journals share some common trajectories of publication. For example, in both the EJSP and the JPSP, studies on aggression conducted mainly via experiments were predominant in the first decades and later decline, although, in the same periods the influence of neo-behaviourism is stronger in the JPSP. During the same periods, attitude change and attribution studies (e.g. *dissonance*) were also common features. In the JPSP studies on attribution are also characteristic of the first topic, which contains studies on consensus and relating to social cognition (e.g. *bias, heuristic*), and are also found in the subsequent years (mainly the 1990s). This is true to a lesser extent of the EJSP, where words relating to heuristic contained in the first topic peaked in the 1990s. The attention to culture and politics is very similar both in contents and in trajectory in the two journals (studies on culture are mainly cross-cultural and investigations of politics mainly centre on identifying how different characteristics are typical of one ideology or another), though the journals refer to different contexts. In the JPSP, however, group and intergroup processes are also part of the topic related to culture, though more space is devoted to them in the EJSP, where distinctive topics associated with them are identified. Cognitive processes (in conjunction with the cognitive turn) are also found in both journals (mainly topics two and three, and topic one to some extent, in the EJSP and topics one, two, and three in the JPSP). In both journals, studies relating to memory and categorisation were found to be predominant in the 1990s, and studies of gender and the implicit are more characteristic of the last years considered; in the EJSP, however, there is a specific topic for them, whereas in the JPSP they are more intertwined with studies of personality. Moreover, studies of emotions, motivation and self are afforded more space in the JPSP, while in the EJSP they are found mostly in the fourth topic, where they follow a specific direction and theory (mood and regulatory focus theory). Despite these differences, both topics (topic four in the EJSP and topic three in JPSP) are particularly well represented in the last years included. These findings also confirmed that the two journals reflect the differences between American and European social psychology identified by Scherer (1992), viz., greater attention to personality (as, for example, can be seen from the predominance in the last years of the two topics relating to relationships and well-being and coping) on the one hand, and the attention focused on group and intergroup processes and related issues in the EJSP on the other. The respective distinctions can be found, albeit marginally, in both journals.

4.5 Concluding Remarks

This chapter has highlighted the common and distinctive features found in EJSP and JPSP publications. The two journals showed a number of shared trajectories at their inception (studies on aggression and attribution, among others) and more recently (including studies on gender using implicit measures and culture). However, the distinctive features that characterise American social psychology, viz., the focus on the individual, and the European one, viz., the attention devoted to the social, seem to persist. In the JPSP, this distinction is evident from the space given to personality measurements and the attention to close relationships and individual well-being and coping strategies. In the EJSP, it can mainly be seen from the attention towards social processes and issues, and from the predominance of studies of intergroup dynamics and related social issues, such as religion and migrations.

However, the name itself of the North American journal includes the word "personality", and a specific section devoted to it, denoting a view of social psychology intertwined with personality. Also, it has been noted that most social psychologists in North America also work on personality psychology (Jones 1985), as is confirmed by APA Division 8, the *Society for Personality and Social Psychology*. The question is, how similar would be the publications of the two journals be without considering the personality section? Trends seem to converge as regards the predominance of experimentation and the presence of the social cognition paradigm. In fact, social cognition is extensively covered together with personality in the JPSP, whereas this seems not to be the case for interpersonal relations and intergroup processes, which, like the others, are dealt with in an independently edited section of the journal. An analysis that also takes the section labels used to classify the papers could provide useful insights for a fuller comparison.

Yet other insights could emerge from determining which topics are unexpectedly underrepresented. For example, Scherer (1992) notes that emotions and affects are common interests in both North America and Europe, but this is not reflected by our comparison between the journals. In fact, such topics are more generally mentioned in the JPSP, while in the EJSP they appear only in connection with regulatory focus theory, as indicated above.

As regards the approach used in this study, distant reading—which involves a quantitative analysis of texts mainly performed in an automatic setting—may lead to a loss of detail if compared with a qualitative analysis. On the other hand, it makes it possible to process a larger amount of data than could be considered in a qualitative (close) reading, and this is crucial for a historical view of publication trends, as is our goal here. Thus, in one of the first forays into the growing and largely unexplored field of textual data analysis, we have traced two histories of publications in social psychology, one North American and the other European. These histories, and these publications, are intertwined at many points. It has to be taken into account that we are considering only two, although pivotal, publications. Thus, to consider other sources of analysis, that permit to reflect as well what is missed, could be an added value. Nevertheless, the history we have been able to outline offers a bridge that can contribute to the debate surrounding American and European social psychology.

References

Allport, F. (1924). *Social psychology*. Boston, MA: Houghton Mifflin.

Allport, G. W. (1954). The historical background of modern social psychology. In G. Lindzey (Ed.), *Handbook of social psychology* (Vol. 1, pp. 3–56). Cambridge, MA: Addison-Wesley.

Allport, G. W. (1968). The historical background of modern social psychology. In G. Lindzey & E. Aronson (Eds.), *Handbook of social psychology* (Vol. 1, 2nd ed., pp. 1–80). Reading, MA: Addison-Wesley.

Allport, G. W. (1985). The historical background of modern social psychology. In G. Lindzey & E. Aronson (Eds.), *Handbook of social psychology* (Vol. 1, 3rd ed., pp. 1–46). Reading, MA: Addison-Wesley.

Baumeister, R. F., & Vohs, K. D. (2007). *Encyclopedia of social psychology*. Thousand Oaks, CA: Sage.

Bolasco, S. (2013). *L'analisi automatica dei testi: fare ricerca con il text mining*. Roma: Carocci.

Contarello, A., & Mazzara, B. M. (2000). *Le dimensioni sociali dei processi psicologici: individui, contesti, appartenenze*. Roma-Bari: Laterza.

Danziger, K. (1995). Neither science nor history? *Psychological Inquiry, 6*(2), 115–117.

Farr, R. M. (1996). *The roots of modern social psychology, 1872–1954*. Oxford: Blackwell Publishing.

Fisch, R., & Daniel, H. D. (1982). Research and publication trends in experimental social psychology: 1971–1980—a thematic analysis of the Journal of Experimental Social Psychology, the European Journal of Social Psychology, and the Zeifschrift für Sozialpsychologie. *European Journal of Social Psychology, 12*(4), 395–412.

Fiske, S., & Taylor, S. E. (1982). *Social Cognition*. New York: Random House.

Graumann, C. F. (1995/1999). *The origins of the EAESP: Social psychology in Europe*. Retrieved from http://www.easp.eu/history/?

Jaspars, J. (1986). Forum and focus: A personal view of European social psychology. *European Journal of Social Psychology, 16*(1), 3–15.

Jones, E. E. (1985). Major developments in social psychology during the past five decades. In G. Lindzey & E. Aronson (Eds.), *The Handbook of Social Psychology* (3rd ed., Vol. 1, pp. 47–107). New York: Random House.

Katz, D. (1965). Editorial. *Journal of Personality and Social Psychology, 1*(1), 1–2.

Katz, D. (1966). Editorial. *Journal of Personality and Social Psychology, 3*(1), 1–2.

Kruglanski, A. W., & Stroebe, W. (2012). The making of social psychology. In A. W. Kruglanski & W. Stroebe (Eds.), *Handbook of the history of social psychology* (pp. 3–17). New York: Psychology Press.

Lebart, L., Salem, A., & Berry, L. (1998). *Exploring textual data*. Dordrecht: Kluwer Academic Publishers.

Lubek, I., & Apfelbaum, E. (2000). A critical gaze and wistful glance at handbook histories of social psychology: Did the successive accounts by Gordon Allport and successors historiographically succeed? *Journal of the History of the Behavioral Sciences, 36*(4), 405–428.
Manstead, A. S., Hewstone, M. E., Fiske, S. T., Hogg, M. A., Reis, H. T., & Semin, G. R. (1995). *The Blackwell Encyclopedia of social psychology*. Oxford: Blackwell Reference/Blackwell Publishers.
Moretti, F. (2013). *Distant reading*. London, UK: Verso/New Left Books.
Moscovici, S. (1999). Ringraziamento. In *Laurea Honoris Causa in Psicologia a Serge Moscovici*. Università degli studi di Roma "La Sapienza": Centro Stampa d'Ateneo.
Moscovici, S., & Markova, I. (2006). *The making of modern social psychology*. Cambridge, MA: Polity.
Palmonari, A., & Emiliani, F. (2014). Introduzione: La teoria delle rappresentazioni sociali nell'evoluzione della psicologia sociale. In A. Palmonari & F. Emiliani (Eds.), *Psicologia delle rappresentazioni sociali. Teoria e applicazioni* (pp. 7–40). Bologna, Il Mulino.
Ratinaud, P. (2014). Visualisation chronologique des analyses ALCESTE: application à Twitter avec l'exemple du hashtag# mariagepourtous. In *Actes des 12es Journées internationales d'Analyse statistique des Données Textuelles*. Paris: Sorbonne Nouvelle–Inalco.
Reinert, M. (1986). Un logiciel d'analyse lexicale: Alceste. *Les Cahiers d'Analyse des Données, 9*(4).
Reinert, M. (1993). Les «mondes lexicaux» et leur «logique» àtravers l'analyse statistique d'un corpus de récits de cauchemars. *Langage et Societe, 66*, 5–39.
Ross, L., Lepper, M., & Ward, A. (2010). History of social psychology: Insights, challenges, and contributions to theory and application. In S. T. Fiske, D. T. Gilbert, & G. Lindzey (Eds.), *Handbook of social psychology* (Vol. 1, 5th ed., pp. 3–50). Hoboken, NJ: Wiley.
Sbalchiero, S., & Tuzzi, A. (2016). Scientists' spirituality in scientists' words. Assessing and enriching the results of a qualitative analysis of in-depth interviews by means of quantitative approaches. *Quality and Quantity, 50*(3), 1333–1348.
Scherer, K. R. (1992). Social psychology evolving. A progress report. In M. Dierkes & B. Biervert (Eds.), *European social science in transition: assessment and outlook* (pp. 178–243). Campus, Westview, Frankfurt: Boulder.
Scherer, K. R. (1993). Two faces of social psychology: European and North American perspectives. *Social Science Information, 32*(4), 515–552.
Tesser, A. (1991). Editorial. *Journal of Personality and Social Psychology, 61*(3), 349–350.
Trevisani, M., & Tuzzi, A. (2015). A portrait of JASA: The history of Statistics through analysis of keyword counts in an early scientific journal. *Quality and Quantity, 49*, 1287–1304.
Trevisani, M., & Tuzzi, A. (2018). Learning the evolution of disciplines from scientific literature. A functional clustering approach to normalized keyword count trajectories. *Knowledge-Based Systems, 146*, 129–141.
Tuzzi, A. (2003). *L'analisi del contenuto: introduzione ai metodi e alle tecniche di ricerca*. Roma: Carrocci.

Chapter 5
First Steps in Shaping the History of Linguistics in Italy: The Archivio Glottologico Italiano

Giovanni Urraci and Michele A. Cortelazzo

Contents

5.1	Introduction	88
5.2	The Corpus	88
5.3	Analysis Methods	90
5.4	Correspondence Analysis	91
5.5	The Incidence of Technical Terms in the Vocabulary of the Journal	97
5.6	Discussion and Conclusions	99
References		102

Abstract This chapter presents the preliminary results of a study on the history of linguistics in Italy. Our purpose is to trace the evolution of the discipline analysing, by means of quantitative methods, the changes in its terminology. To achieve that, we examined the oldest Italian linguistics journal, the *Archivio Glottologico Italiano*, started in 1873 and still published today.

Using correspondence analysis, it was possible to split the history of the journal into six chronologically compact periods and to associate each of them with a specific set of representative terms: every phase has a different lexical profile, and the changes in the vocabulary can be interpreted with the support of extratextual information. Two periods (1910–1923 and 1989–1993) are particularly important: they do not correspond to specific moments that mark the history of the journal; however, text analysis shows that they have their own strong identity and that they played an important role in the evolution of the journal. Furthermore, we have investigated the amount of technical terms in each issue of the journal: while the size of the technical vocabulary has proved to be essentially consistent throughout the years, the relative frequency of the linguistic terminology shows a steep increase at the end of the 1980s.

G. Urraci (✉)
University "Ca' Foscari" Venice, Venice, Italy
e-mail: giovanni.urraci@unive.it

M. A. Cortelazzo
University of Padova, Padova, Italy

A. Tuzzi (ed.), *Tracing the Life Cycle of Ideas in the Humanities and Social Sciences*, Quantitative Methods in the Humanities and Social Sciences,
https://doi.org/10.1007/978-3-319-97064-6_5

Keywords Corpus linguistics · History of linguistics · Archivio Glottologico Italiano · Correspondence analysis · Technical terms

5.1 Introduction

The birth of linguistics as a science in Italy is usually associated with Graziadio Isaia Ascoli. He was professor of comparative grammar and Eastern languages since 1861 at the *Accademia scientifico-letteraria* that later became the University of Milan (Cortelazzo 1973; Benincà 1994; Marazzini 2011; Morgana 2010; Lubello 2016). In 1873, Ascoli founded, with Giovanni Flechia, the first Italian journal of linguistics, the *Archivio Glottologico Italiano*, still published today, albeit some interruptions and a long break between 1942 and 1950. In 2015, the journal reached its hundredth volume. Graziadio Isaia Ascoli promoted systematic linguistic research, especially while being chief editor of the *Archivio Glottologico Italiano*. He also promoted research on the Italian languages (Italian and dialects), including the modern use of language, ancient dialects, and written documents.

The *Archivio Glottologico Italiano* is therefore the oldest linguistics journal published in Italy. It was directed by some of the most important linguists in Italy (Carlo Salvioni, Pier Gabriele Goidanich, Matteo Bartoli, Vittore Pisani, Benvenuto Terracini, Giacomo Devoto, Carlo Alberto Mastrelli, Romano Lazzeroni). In 2018, it is being directed by Alberto Nocentini and Paolo Ramat. This is why the *Achivio Glottologico Italiano* is the perfect example to represent the evolution of Italian linguistic studies analysed from an internal point of view, meaning through the vocabulary of the articles and the most representative technical terms. Our approach is different from those used in other overviews of the history of linguistics in Italy (Ramat et al. 1986; Benincà 1994): the object of our study is not the theoretical approach of the main Italian linguists, but the actual research practice that emerges from the articles published on the journal, with Trevisani and Tuzzi (2015, 2018) as a methodological reference model.

This study is therefore based on an analysis "from afar" (*distant reading*: Moretti 2005) of the vocabulary of the journal. The analysis on the lexical evolution of the *Archivio Glottologico Italiano* is still a work in progress: it will later analyse other Italian linguistics journals using several methodological approaches, both quantitative and qualitative. Here, we will give an overall view that is, however, the premise of a diachronic study of the Italian linguistics terminology (see De Felice 1954 for the first phase, and De Luca 2014 for the period between the 1940s and the end of the 1970s).

5.2 The Corpus

The corpus is composed by all the articles of the journal, starting from issue no. 2 published in 1876 to issue no. 99 published in 2014. The issue published during the first year (1873) was excluded since it only consists of an essay by Graziadio Isaia

Ascoli (*Saggi ladini*) and can therefore be considered a monographic issue. Volumes 5 and 6 were also excluded, as they contain the text and the critical edition of an ancient code. Furthermore, year 22 and year 23 are merged into a single volume.

Creating the corpus has not been easy. Indeed, either digital versions of the years were not available or the existing ones did not guarantee correctness. The corpus has been created following the standard digitalization process. The texts were scanned, then processed with an OCR, then checked automatically by using a macro to correct the most frequent errors, but also by reading it ourselves. We decided to work on whole texts mainly to systematically consider all their contents. However, it was also a necessary choice, as abstracts are used only since 2000 and the titles of the articles rarely proved to be informative and representative of all the lines of research.

The digitalized texts were later analysed according to their intrinsic characteristics. First of all, the texts published in a language other than Italian have been excluded (they are more frequent in recent years, as the journal has increasingly received articles in English, but in the past there have been articles published in other languages: there are 52 articles in foreign languages, of which 33 published since 1992; French, Spanish, German, and Portuguese are used besides English). We also excluded articles containing lexica of dialects, specialized glossaries, editions of ancient texts (e.g. the article by M. Gaster, *La versione rumena del Vangelo di S. Matteo, tratta dal Tetraevangelion del 1574,* issued in volume no. 12, pp. 196–254), as well as parts of articles with said features. We have also left out additions, corrections, and annotations; references, news records, reviews, obituaries, analytical indexes, and lists of received publications, graphs and tables, examples, chunks of texts and quotes in languages other than Italian if longer than two lines, and all the abstracts, a feature introduced in recent years. We therefore succeeded in getting a compact and coherent corpus, made up of sub-corpora as homogeneous as possible to make the comparison between years more effective. Furthermore, excluding texts and quotes in dialects and other languages allowed us to build a homogeneous corpus with a real focus on the language of scholars, not on the testimonies of different times and places. These dialectal elements also made the corpus spurious, as demonstrated by the high percentage of hapax (about 80% of the vocabulary) in the original version of the corpus.

The text was then normalized: we wrote abbreviations in full, changed all letters to lowercase, homogenized the spelling of compound words that were written with a hyphen (e.g. *neo-grammatico* > *neogrammatico*, *fono-sintattico* > *fonosintattico*, *mediopassivo* > *medio-passivo*), wrote apocopated words in full (e.g. *vocal* > *vocale*), normalized words with prosthetic vowels (e.g. *istrumentale* > *strumentale*), chose *i*s instead of *j*s in words with a semivowel *i* (e.g. *conjugazione* and *jato* now become *coniugazione* and *iato*). Finally, we homogenized words with orthographic or phonetic variants (e.g. *indeuropeo* > *indoeuropeo*, *napolitano* > *napoletano*).

After these processes, the corpus is composed of 831 articles from 95 volumes of the journal.

We report in Table 5.1 the lexical measures of the corpus.

Table 5.1 Basic lexical measures of the corpus

N—word-token	5,089,527
V—word-types	268,920
(V/N)*100—type/token ratio	5.3
(V_1/V)*100—percentage of hapax	61.8

5.3 Analysis Methods

The corpus was analysed using quantitative text analysis tools. First, considering the features of the corpus, we tried to find words that could be considered technical terms. A reference lexical list of entries was created choosing from nine Italian terminology dictionaries (Beccaria 2004; Cardona 1969, 1988; Casadei 2011; De Felice 1954; Dubois et al. 1979; Ducrot and Todorov 1972; Gentile 1963; Severino 1937). We added to these lists a set of relevant entries manually extracted from the corpus (especially entries that have changed their meaning over time and words now marginal that used to be significant technical terms). A total of 7950 specialized terms was collected this way. They include 3019 multiword expressions, 448 names of languages and dialects, 357 foreign words, and 694 technical terms from other disciplines often used in linguistics. The corpus contained 4995 technical terms of our list (63%), meaning that the Archivio Glottologico Italiano covers a good portion of the vocabulary of linguistics, and therefore of its related topics.

At an operational level, this list was also used to create three categorical variables using the vocabulary of the corpus. They define each lexical unit as technical term (yes/no), name of language (yes/no), and technical term from other discipline (yes/no). These variables allowed us to find the relevant vocabulary on which to focus on the following analyses. Furthermore, the list of technical terms was important to find multiword expressions (MWE) and was used to produce a list of lexicalizations to be used in the processing of the corpus with the TalTac2 software (Bolasco 2010). However, not all the detected MWEs have been used, but only some of them to avoid dispersing occurrences when dividing them in more units of analysis. Therefore, we lexicalized all forms with more than 15 occurrences, we excluded those with less than 8 occurrences, whereas for those forms with 8–15 occurrences we only kept the multiword expressions with disambiguating value and those not competing with simple forms. Overall, the MWEs considered are 576. All of them are monosemic and with a high degree of specialization.

At least in this phase, we worked on forms instead of lemmas. Indeed, automatic lemmatization in Italian is still uncertain and insufficient for texts with many technical terms.

Until now, we have used two methods of text quantitative analysis. The first is given by correspondence analysis (see Greenacre 1984, 2007; Murtagh 2005, 2010; Lebart et al. 1984, 1998). It gives a general representation of the time pattern of the journal: projecting years and words on a plane shows similarities among the issues of the journal and among the selected keywords, associating a list of words to a specific time period. Our calculations were based on a contingency table reporting keywords × year and containing 8860 forms, selected on the basis of a double frequency threshold (>49 for common words, >8 for technical terms). The three dichotomous variables associated with each keyword previously described (technical term, name of language, technical term from other discipline) have been used to improve the readability of the visualization of words on the graphs. The results of the analysis are shown in Sect. 5.4. The second analysis technique allows to consider changes of the specialized vocabulary over time. It consists in calculating two measures in the different time periods of the journal: the size of the technical vocabulary (types per years), and the ratio between technical terms (tokens) and the sub-corpus dimension (meant as texts of the same year). The data obtained provide an overall perspective of the changes in the technical vocabulary in terms of weight, density, and distribution. The results of the second analysis are shown in Sect. 5.5.

We are also currently working on topics analysis using Reinert's method (with the Iramuteq software) and the projection of the topics on the time axis (with the R software) (see Ratinaud 2014 and Chap. 10 in this volume). However, this analysis is still ongoing and we will discuss the results in the future.

5.4 Correspondence Analysis

Correspondence analysis was used as the main technique to analyse the vocabulary of the *Archivio Glottologico Italiano* over time. The 95 volumes of the journal are represented on the first factorial plane (the first two axes out of a total of 94) with an explained inertia percentage of 11.51% (7.17% for the first one, 4.34% for the second one). The representation of the results for the first and second factor is shown in Fig. 5.1.

The correspondence analysis shows a regular time distribution of the volumes of the journal, which group in six clusters that correspond to as many time periods. Four out of the six clusters correspond to the main evolution periods of the contents of the journal, whereas the remaining two can be identified as transitory phases that precede relevant changes.

The upper-left quadrant shows the years since the foundation until the beginning of the 1920s; the bottom-left shows those from the 1920s to the 1950s; the bottom-right shows those from the 1950s to the mid-1990s; and the upper-right shows those since the second half of the 1990s until today.

Before examining each quadrant, some general characteristics may be observed, also with the help of Fig. 5.2.

The most evident observation is that the upper-left and the upper-right quadrant present a higher number of technical terms. The upper-left quadrant contains a high number of peculiar terms, with many names of languages and dialects. In the upper-right quadrant, which collects the most recent years, there is a high percentage of characterized vocabulary that tends to cluster. This shows an increasing similarity between the issues of the journal, with a steady presence of lexical innovations and

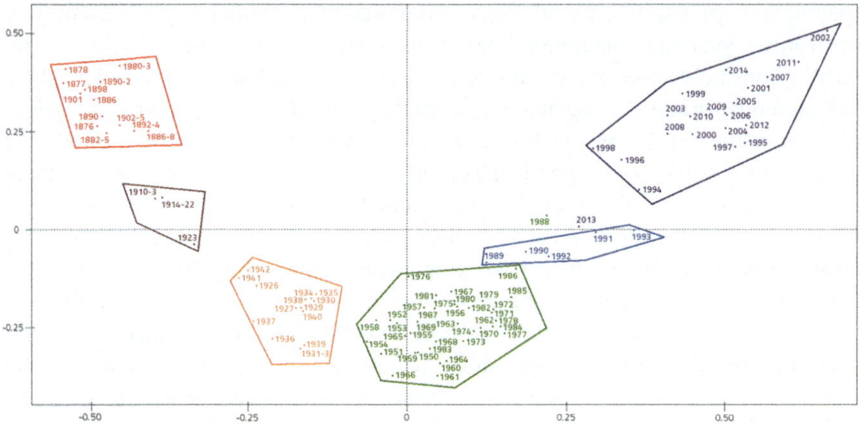

Fig. 5.1 First factorial plane of correspondence analysis. Projection of years manually divided into six clusters

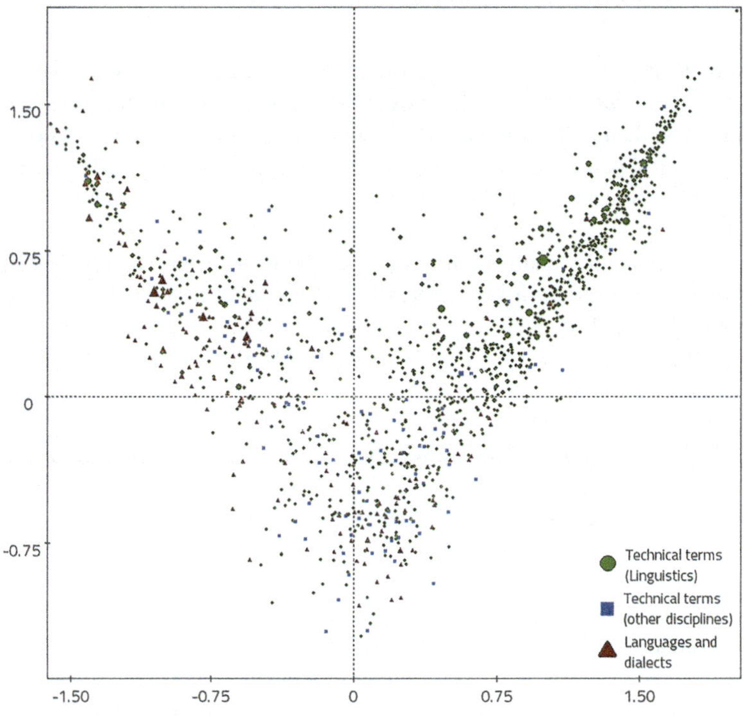

Fig. 5.2 First factorial plane of correspondence analysis, projection of 50% of the keywords with the largest contribution. The size of the symbol is proportional to the contribution of the word

a higher internal homogeneity that correspond to a decrease in the number of subjects. In short, the fourth period is characterized by few newer topics with a high degree of specialization. The bottom-left quadrant seems to correspond to a transition period: It does not have a strong identity and its vocabulary shares many features with the other periods. Finally, the bottom-right quadrant shows a significant presence of terms from other disciplines. This corresponds to the new topics and methods found in the corresponding years.

A detailed characterization of the periods taking into account period-specific words (nouns, adjectives, verbs) can be obtained from a closer look at each quadrant.

We selected the words to be projected on each quadrant in two phases. First, we classified the vocabulary according to the categorical variables already discussed in Sect. 5.3 (technical terms of linguistics, names of languages, technical terms of other disciplines). We examined only the technical terms of linguistics and those of other disciplines. Indeed, we believe that only these two categories can describe the evolution of the linguistic thought in the journal. Therefore, the following figures do not show names of languages and common words. We later applied to this selected vocabulary a filter based on the contribution given by single words to the determination of the year coordinates. To obtain both informative capacity and readability, we set a different threshold for each quadrant. The upper-left quadrant shows the top 25% of the technical terms with the highest contribution, the bottom-left and bottom-right quadrants show the top 40%, and the upper-right quadrant the top 10%.

The upper-left quadrant, shown in Fig. 5.3, contains the specific words of the first period of the journal, since its origins to the 1920s (however, the years 1910–1923 appear as a transition period to the next phase, contained in the bottom-left quadrant).

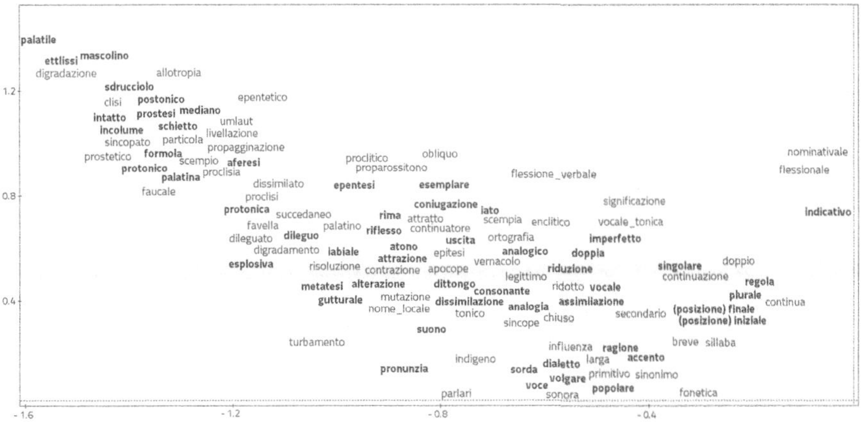

Fig. 5.3 First factorial plane of correspondence analysis. Projection on the upper-left quadrant of 25% of the keywords with the highest contribution (top 10% marked in bold)

The upper-left quadrant is characterized by a substantial set of specialized terms. Their strong affinity suggests that methodologies and interests are homogeneous. The main feature of the upper-left quadrant is its high number of technical terms related to phonetics, besides some lexical clues that refer to a substantially neogrammarian background. Among the most distinctive terms of this first macro-period we can find:

1. General terms as *allotropia, cimelio, continuatore, incolume, intatto, legittimo, primitivo, secondario, succedaneo, risoluzione*, with many references to analogy, one of the primary forces of linguistic change according to neogrammatics: *analogia, analogico, livellazione.*
2. Terms related to phonetics. These include generic terms as *vocale, consonante, dittongo, iato, accento, atono, postonico, protonico, tonico*; sound names, especially consonant-related terms (places and manners of articulation) as *esplosiva, gutturale labiale, palatile, palatina, sonora, sorda*; names of generic phonetics processes: *aferesi, assimilazione, digradazione, dileguo, dissimilazione, epentesi, ettlissi, metatesi, prostesi, riduzione.*
3. Names of languages and dialects: words as *dialetto, favella, parlari, vernacolo,* and adjectives as *popolare* and *volgare.*

In summary, the upper-left quadrant shows a good lexical specificity, consisting mainly in a compact set of terms that refers to phonetic studies researching diachronic changes.

The bottom-left quadrant is shown in Fig. 5.4.

The bottom-left quadrant does not contain many technical terms. Indeed, the vocabulary in this period follows the trends of the previous period or anticipates trends that came later. However, the vocabulary of this period revolves around the nucleus of neolinguistics (also known as spatial linguistics).

The words in this quadrant can be grouped in different clusters too:

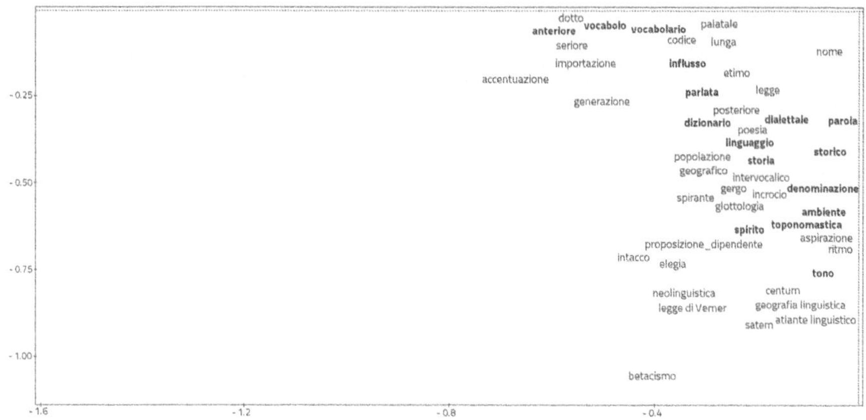

Fig. 5.4 First factorial plane of correspondence analysis. Projection on the bottom-left quadrant of 40% of the keywords with the highest contribution (top 20% marked in bold)

1. Terminology related to neolinguistics and its tools: first of all, *neolinguistica*, *geografia linguistica*, *atlante linguistico*, *geografico*, but also *vocabolario* and *dizionario* and *storia*, *storico*, *etimo*; besides the name of the discipline in that period: *glottologia*.
2. Names of the researched topics: from *linguaggio* (that replaces more specific names, such as *dialetto, favella, parlari, vernacolo,* used in the previous period; in particular, Bartoli used *linguaggio* as a hypernym of *lingua* and *dialetto* (Graffi 2010, p. 172)) to *denominazione*, *nome*, *parola*, *vocabolo*. Thus, it is clear more attention is paid to the vocabulary compared to the previous period, more focused on phonetics.
3. Typical ideas of spatial linguistics, such as *anteriore*, *seriore*, *imitazione*, *importazione*, *and incrocio.*
4. Expressions like *legge*, especially in *legge di Verner* (an indicator of the still strong influence of the research carried out by neogrammatics) or (*lingue*) *centum*/*satem* (typical expressions taken from Indo-European studies).

In summary, the bottom-left quadrant mainly revolves around the nucleus of neolinguistic terminology and its new ideas, as regards both methods and objects of analysis.

The technical terms in the bottom-right quadrant (Fig. 5.5) do not show a strong distinctive identity. They reiterate previous interests or reveal new approaches, disciplines, and theories that do not prevail each other. The result shows a high lexical richness with an exceptional variety of references. Also as for this quadrant, we can find several relevant word categories:

1. Words that refer to currents of linguistics or complementary disciplines: *dialettologia*, *grammatica*, *linguistica*, *linguista*, *lingua*, *linguistica storica*, *psicologia*, *retorica*, *sociolinguistica*, *stilistica* (with *stile*, *stilistico*, *prosa*, *poetica*, *unità melodica*), *storia della lingua*, *storia linguistica*, *strutturalismo*.

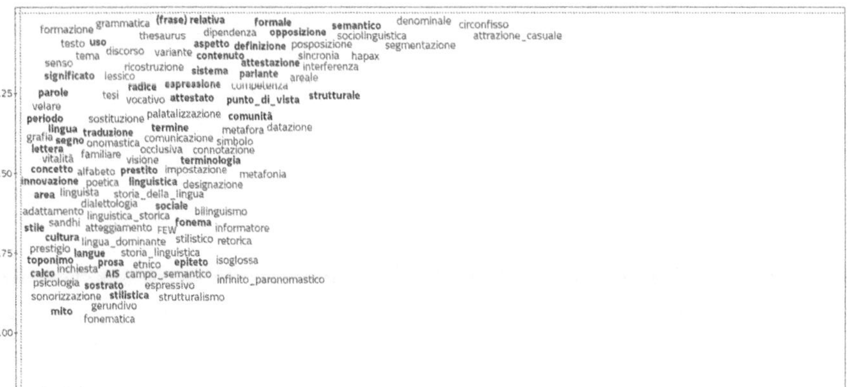

Fig. 5.5 First factorial plane of correspondence analysis. Projection on the bottom-right quadrant of 40% of the keywords with the highest contribution (top 20% marked in bold)

2. Words that refer to structuralism, mainly to the Saussurian thought: *langue, parole, segno, significato, sincronia, strutturale*, sistema.
3. Words that refer to a sociolinguistic perspective (*bilinguismo, comunicazione, comunità, cultura, lingua, dominante, prestigio, registro familiare, sociale*) or to an interest in contact phenomena (*adattamento, calco, interferenza, prestito*). We also count in this categories words as *competenza* e *parlante*, both shared by sociolinguistics and transformational-generative linguistics. However, these should be linked primarily to sociolinguistics studies in these years of the *Archivio Glottologico Italiano*.
4. Vocabulary related to historical linguistics (*attestazione, datazione, ricostruzione, sostrato*), still one of the main topics of the journal in that period.
5. Words that refer to phonetics (*metafonia, monottongazione, occlusiva, palatalizzazione, sandhi, sonorizzazione, velare*) and, to a much lesser degree, to phonology, which shows a very limited impact (*fonema, fonematica, opposizione*, especially referring to phonological opposition).
6. Words that refer to morphology and syntax, anticipating the most relevant feature of the upper-right quadrant: *aspetto, attrazione casuale, circonfisso, denominale, dipendenza (frase)relativa, posposizione*. On the other hand, terms referring to semantics are underutilized. In fact, we can only report *campo semantico, connotazione, semantico*.

In summary, the bottom-right quadrant is characterized by a great lexical variety that suggests new interests deriving from theoretical and methodological innovations, but also from the openness towards other disciplines. We can therefore conclude that the structuralist theories were not influential enough to impose a new line of research. Indeed, although the Italian linguists generally acknowledged the value of De Saussurre's theories, the cultural interpretation of language still prevailed over the interest for the internal relations of the language (Segre 1986, p. 261). This clearly emerges from the vocabulary used in the journal.

So, we get to the last quadrant, represented in Fig. 5.6.

The upper-right quadrant is characterized by a strong compactness that is not comparable to the dispersion in the other quadrants. Indeed, the significant words in this phase are all associated with morphology and syntax, with a large number of words related to this area of linguistic research; this concerns, for example, the names of verb moods and tenses (*aoristo, condizionale, congiuntivo, gerundio, imperativo, participio, presente*), names of parts of speech (*avverbio, congiunzione, pronome, verbo*), but also words related to newer concepts, such as *valenza, inaccusatività, marcato, ruolo tematico, verbo sintagmatico, costituente*, and *determinante*. This is a major innovation in the history of the journal since morphology and syntax had played a secondary role until then.

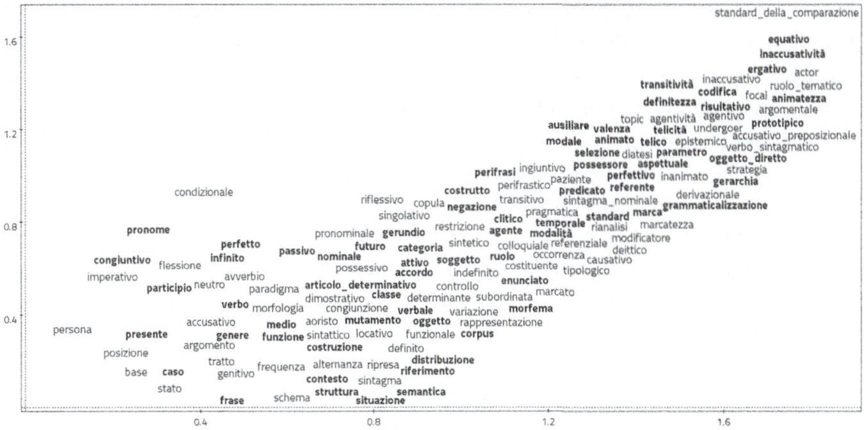

Fig. 5.6 First factorial plane of correspondence analysis. Projection on the upper-right quadrant of 10% of the keywords with the highest contribution (top 5% marked in bold)

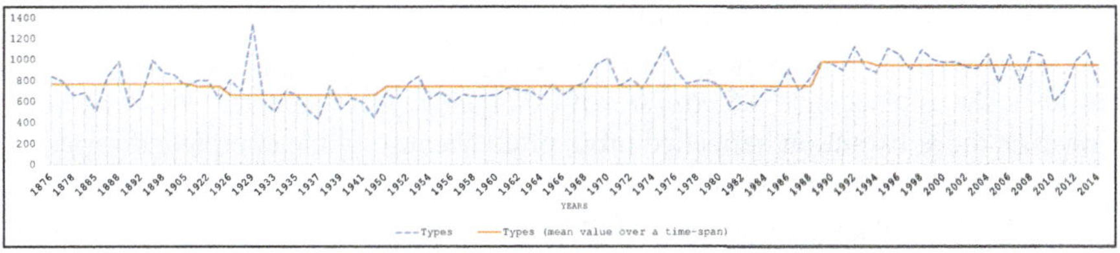

Fig. 5.7 Evolution over time of the dimension of the specialized vocabulary

5.5 The Incidence of Technical Terms in the Vocabulary of the Journal

We have observed the diachronic evolution of the technical terms considered as a whole and observed from a quantitative point of view. Our observation had two goals: the first was describing the variations in the dimension of the specialized vocabulary. The second one was measuring the frequency with which the technical terms occur in the texts (a data that can define the degree of specialization of the discourse). All our considerations will be based on the data represented in two graphs: the first shows the number of technical terms (types) used per year and the average value for each of the periods identified by correspondence analysis; the second summarizes the relative frequencies, showing them both as a per year value and as an average value during the period.

In Fig. 5.7 is therefore shown the number of different technical terms (types) in each issue of the journal. It also shows the dimension of the specialized vocabulary of the articles. Absolute values are reported, as they represent a simple but also effective measure for evaluating the dimensions of the specialized vocabulary. They also represent a good basis for comparing years. Indeed, the size of the articles, and consequently their vocabulary, affects the tokens, but their influence is negligible on the types because the number of different technical terms needed to discuss a topic is steady and hardly varies in proportion to the length of the text.

The number of technical terms per year is substantially uniform throughout most of the history of the *Archivio Glottologico Italiano*, excluding a slight decrease in the years between the two world wars and a steeper decrease in 2010–2011 (due to special monothematic issues dedicated to grammaticalization). However, in the last 25 years, since the end of the 1980s, there is a sudden peak instead of a slow progression, with 1989 as a turning point. Yet, the observed increase is smaller than what could have been expected: in almost 150 years, starting from a discipline still in its early stages, the size of the technical vocabulary used annually increases only by 22%, with average values that, however, are not far from those recorded in several previous years of the journal. This suggests that the vocabulary, and therefore the linguistic knowledge, does not grow incrementally, by accumulation, but by substitution. A substitution which, to a lesser extent, takes place by replacing some terms with more recent synonyms related to the new points of view of the discipline (a prime example of this is the case of *colore*, completely replaced by *articolazione*; but also, the replacement of *ettlissi* with *dileguo*, and of *accatto* with *prestito*). However, the substitution happened more often because of the complete disappearing of some research topics and ideas, and because of the introduction of new topics and models carrying a substantial quantity of new terms as well as already-existing words with a new technical meaning (this is what happened in recent years with *strategia* and *tema*, whereas *chiaro* and *(o)scuro* went through the opposite process and lost their technical meaning referred to vowels).

Figure 5.8 shows, year by year, the ratio between the technical term tokens and the length of the sub-corpus (multiplied by 1000). The graph therefore represents the degree of coverage of the text by the technical terms, a significant sign of the degree of specialization of the discourse.

The curves in the graph show a clear and strong increasing trend, and, compared to the graph shown in Fig. 5.7, at the end of the 1980s an even more noticeable step takes place: the gap between the first and last period is remarkable, with a growth of 65% in the frequency of specialized vocabulary—the percentage growth of tokens is three times higher than the one recorded in types. In other words: the speech tends to become denser and more loaded with technical terms. To explain this phenome-

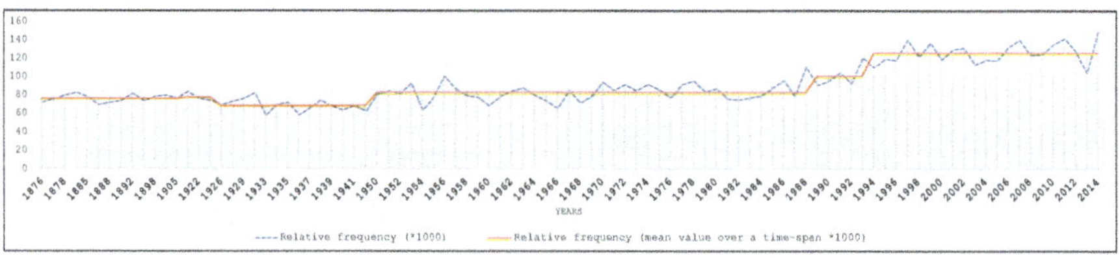

Fig. 5.8 Evolution over time of the relative frequency of technical terms

non, a detailed qualitative analysis will be needed. As of now, we can however assume that it depends on the introduction of research topics characterized by a higher degree of specialization that therefore uses more linguistic technical terms, as already appeared from the correspondence analysis presented in the previous section.

5.6 Discussion and Conclusions

The results of the study on the *Archivio Glottologico Italiano* look promising at this early stage.

The correspondence analysis shows a clear division of the journal into six groups, referring to six periods. Although only one factorial plane was taken into account, just 2 years, 1988 and 2013, do not follow the chronological order. Regarding the year 2013, the reason is quite clear: that issue was published in memory of a scholar, Alberto Zamboni, who belonged to the previous period. His colleagues reasonably decided to honour his memory retracing his steps and outlining his methodologies and lines of research. Which leaves us with just the year 1988, slightly distant from the contiguous years.

During the first period (1876–1905), the chief editor was Graziadio Isaia Ascoli, together with co-founder Giovanni Flechia, followed by Carlo Salvioni. This period is rather homogeneous from a lexical point of view. On the background, which consists of terms belonging to the early stages of scientific linguistics (the Neogrammarians), names of general phenomena can be found, thanks to which the evolution of language stops being a mere mechanical application of phonetic laws. Dialects were the main focus during this period, thanks to the influence of Giovanni Flechia (Polimeni 2014). The correspondence analysis places the period running from 1910 to 1923 in the same quadrant. Only three volumes belong to this period: it was a transition phase, under chief editor Pier Gabriele Goidanich, who stated in the 1923 issue preface that his objective was to combine dialectology and literary language and its link with dialects (Proietti 2001): as appears from the titles of the issues published during this period, the project was not accomplished. Nevertheless, the intention to innovate led to a strong lexical distinctiviness: the journal transcends the mindset carried out by Ascoli and Salvioni embracing innovation, as appears in the bottom-left quadrant. Goidanich, although averse to spatial linguistics, shows a pluralism of interests and an open-mindedness which will lead to him accepting new models, such as the *Wörter und Sachen* line of research (Malkiel 1986, pp. 284–285): he certainly was the ideal chief editor during this transition phase.

The bottom-left quadrant is populated by issues running from 1926 to 1942, corresponding to the new series of the journal, led by Matteo Bartoli and Benvenuto Terracini. The new period differs substantially from the previous one. Neolinguistics outstrips neogrammarians, as appears also from the fact that the essay *Introduzione alla neolingustica* (translation: *Introduction to neolinguistics*), by Matteo

Bartoli, was described by the author himself as being programmatic for the Neo-Latin section of the *Archivio Glottologico Italiano*. During this period, attention is payed to spatial linguistics (and regarding Italian, to the *Atlante Linguistico Italiano*). It is quite noteworthy that attention is also payed to real-life language, in line with field research and with the idea of language as a social product. Subsequently, during these years, common words can also be significant to the linguistic discourse carried out by the journal (that's the case with words such as *ambiente*, *generazione*, *popolazione* and with plentiful words referring to material culture, to which importance is given thanks to the *Wörter und Sachen* linguistics school).

The bottom-right quadrant is populated by the issues running from 1950 to 1988. It was a period of transformation and experimentation, which seems suitable for global considerations rather than considerations on its internal evolution. During those years innovation, new theories, and various perspectives arise (later in Italy than they did abroad: primarily, structural linguistics). Several chief editors take over, and no perspective prevails: chief editors Benvenuto Terracini, Carlo Mastrelli and Vittore Pisani work alongside prominent linguists such as Giacomo Devoto, Bruno Migliorini and Giuseppe Vidossi. Dialectology, traditionally of interest for the journal, is addressed together with general linguistics, theoretical reflections, comparative linguistics, the protohistory of Italian (the history of Italian language was addressed instead by the journal *Lingua nostra*, founded in 1939 by Bruno Migliorini and Giacomo Devoto). Nevertheless, lexical signals (such as *area*, *informatore*, *inchiesta*, *isoglossa*) show that great attention is payed to linguistic context, in continuity with the previous quadrant. An evolution pattern emerges, prompted by the work on linguistics atlases with more and more attention being payed to sociolinguistics at a time when it was difficult to develop structural descriptions of dialects, leading to dialectologists focusing on data collection methods rather than data analysis (Benincà 1994, p. 612). The bottom-right quadrant also features the period running from 1989 to 1993: a transition phase, marked by the entrance into the journal board of members of the two major professional linguistics association, SLI (*Società di Linguistica Italiana*) and SIG (*Società Italiana di Glottologia*). The members were Tullio De Mauro for SLI, Paolo Ramat for SIG. A more systematic approach to the topics discussed in the journal was adopted, as appears from the increasing number of technical terms, despite the decrease in topics which emerges from CA. During this period, the vocabulary becomes more and more specialized (and also semantically more compact).

The last quadrant features the two most recent decades, 1994–2014. The vocabulary is further characterized, in particular, the specialized vocabulary. The vocabulary also confirms the point made in the preface of issue no. 79, 1994: linguistics is too diversified, and a single journal cannot possibly manage all research fields. The *Archivio Glottologico Italiano* opts for historical linguistics and comparative linguistics. It also adopts a specific research method although not explicitly: as appears from the technical vocabulary, morphosyntactic analysis is carried out. Technical terms grow in number, and focus more and more on morphosyntax. Therefore, words popular in more recent research can be found (such as *agentività*, *animatezza*, *definitezza*, *grammaticalizzazione*, *inaccusatività*, *marcatezza*, *valenza*), as well as words which were traditionally used in grammar studies (e.g. names of parts of speech or verbal tenses and moods) but have not been significantly used in the previous periods because of the predominance of phonetic and lexicographic studies. In contemporary linguistics, a battle of function against form occurred (Graffi 2010, p. 440), and in the *Archivio Glottologico Italiano* the battle was won by the latter, although the trends in the vocabulary of the bottom-right quadrant suggested differently.

What emerges from this paper is that linguistics has undergone many changes in the last 150 years, which occurred as a result of clear fractures brought about by influential, charismatic scholars (in the *Archivio Glottologico Italiano,* the mark of the directors can be effortlessly recognized), and as a result of successive and juxtaposed theories and lines of research. This evolution leads to every period having a clear, recognizable vocabulary (Fig. 5.1), and explains why there cannot be a significant increase in the dimension of specialized vocabulary (Fig. 5.7).

At the same time, a certain degree of continuity emerges, especially in the first four clusters, as appears from homogeneity in the writing (Fig. 5.7 shows coherence in the dimension of the specialized vocabulary; Fig. 5.8 shows stable levels of specialization) and from the persistence of some topics. Although with different perspectives, the first two quadrants are connected by the importance of phonetics and the historical approach, by the ineluctably present neogrammarian approach, and by the charisma and legacy of Graziadio Isaia Ascoli. Between the third and the fourth cluster, the common thread is the evolution of spatial linguistics into sociolinguistics, together with the shift towards the study of morphosyntax: the evolution is carried out through lexical and therefore thematic loans.

After such different yet similar periods, the end of the 1980s is a turning point: the last 25 years exclusively populate the last quadrant of the CA (Fig. 5.1) and are characterized by an unprecedented lexical specificity and thematic cohesion (Figs. 5.2 and 5.6). Specialized vocabulary grows (Fig. 5.7) and technical terms are significantly more present than in the previous texts. The impact of generative grammar leads to a revolution, with more evident consequences on the vocabulary than the ones brought about by structural linguistics. Unsurprisingly, this consolidation occurred after a long period of time (cluster 4, 1950–1988) during which linguistics, as appears from studies in the *Archivio Glottologico Italiano*, had risked losing its identity because of its receptiveness towards other fields (as can be seen from technical vocabulary in Fig. 5.2 and words in Fig. 5.5).

In summary, this work, progressing from the analysis of the use of language, managed to retrace the main steps of the development of linguistics in Italy, endorsing its traditional history. The method used allowed us to focus on the effects of theoretical innovations on research as a practice. The choice allowed us, on the one hand, to associate to each period a specific group of representative words; on the other hand, to measure the distance and juxtaposition of the different periods. The results also led to a new-found importance for two of the periods (1910–1923 and 1989–1993) which, although not a part of the main periodization, have a clear identity and play an essential role into one of the most important Italian linguistics journals. In conclusion, thanks to lexical features, a quick overview was given on how linguistics is discussed in Italian journals, uncovering an unexpected homogeneous trend.

References

Beccaria, G. L. (Ed.). (2004). *Dizionario di linguistica e di filologia, metrica, retorica*. Torino: Einaudi.

Benincà, P. (1994). Linguistica e dialettologia italiana. In G. C. Lepschy (Ed.), *Storia della linguistica* (Vol. III, pp. 525–644). Bologna: il Mulino.

Bolasco, S. (2010). *Taltac2.10. Sviluppi, esperienze ed elementi essenziali di analisi automatica dei testi*. Milano: LED.

Cardona, G. R. (1969). *Linguistica generale*. Roma: Armando.

Cardona, G. R. (1988). *Dizionario di linguistica*. Roma: Armando.

Casadei, F. (2011). *Breve dizionario di linguistica*. Roma: Carocci.

Cortelazzo, M. (1973). *Graziadio Isaia Ascoli e l'Archivio glottologico italiano (1873-1973). Studi raccolti, in occasione del centenario dei Saggi ladini*. Udine: Società filologica friulana.

De Felice, E. (1954). *La terminologia linguistica di G.I. Ascoli e della sua scuola*. Utrecht: Spectrum.

De Luca, M. T. (2014). *Il lessico della linguistica in Lingua nostra (1939-1978)*. Berlin: Logos.

Dubois, J., Mathée, G., Guespin, L., Marcellesi, C., Marcellesi, J.-B., & Mével, J.-P. (1979). *Dizionario di linguistica*. Bologna: Zanichelli.

Ducrot, O., & Todorov, T. (1972). *Dizionario enciclopedico delle scienze del linguaggio*. Milano: Isedi.

Gentile, A. (1963). *Lessico di terminologia linguistica*. Napoli: Liguori.

Graffi, G. (2010). *Due secoli di pensiero linguistico*. Roma: Carocci.

Greenacre, M. J. (1984). *Theory and application of correspondence analysis*. London: Academic Press.

Greenacre, M. J. (2007). *Correspondence analysis in practice*. London: Chapman & Hall.

Lebart, L., Morineau, A., & Warwick, K. M. (1984). *Multivariate descriptive statistical analysis. Correspondence analysis and related techniques for large matrices*. New York: Wiley.

Lebart, L., Salem, A., & Berry, L. (1998). *Ewploring textual data*. Dordrecht: Kluwer Academic.

Lubello, S. (2016). Rapporti italo-tedeschi negli studi linguistici del secondo Ottocento: maestri, scuole, centri culturali. In M. Becker & L. Fesenmeier (Eds.), *Relazioni linguistiche. Strutture, rapporti, genealogie* (pp. 31–49). Frankfurt am Main: Lang.

Malkiel, Y. (1986). Romance and Indo-European linguistics in Italy. In P. Ramat et al. (Eds.), *The history of linguistics in Italy* (pp. 277–299). Amsterdam: John Benjamins.

Marazzini, C. (2011). Storia della linguistica italiana. In R. Simone (Ed.), *Enciclopedia dell'italiano, II* (pp. 1417–1422). Roma: Istituto della Enciclopedia Italiana.

Moretti, F. (2005). *La letteratura vista da lontano*. Torino: Einaudi (in engl., *Distant Reading*, London, Verso, 2013).

Morgana, S. (2010). Ascoli, Graziadio Isaia. In R. Simone (Ed.), *Enciclopedia dell'italiano, I* (pp. 111–113). Roma: Istituto della Enciclopedia Italiana.

Murtagh, F. (2005). *Correspondence analysis and data coding with Java and R*. London: Chapman & Hall.

Murtagh, F. (2010). The correspondence analysis platform for uncovering deep structure in data and information, Sixth Boole Lecture. *Computer Journal, 53*(3), 304–315.

Polimeni, G. (2014). *Il troppo e il vano. Percorsi di formazione linguistica nel secondo Ottocento*. Firenze: Cesati.

Proietti, D. (2001). Goidanich, Pier Gabriele. In *Dizionario biografico degli italiani* (Vol. 57, pp. 558–562). Roma: Istituto dell'Enciclopedia Italiana.

Ramat, P., Niederehe, H.-J., & Koerner, K. (Eds.). (1986). *The history of linguistics in Italy*. Amsterdam: John Benjamins.

Ratinaud, P. (2014). *IRaMuTeQ: Interface de R pour les Analyses Multidimensionnelles de Tewtes et de Questionnaires* [software, Version 0.7 alpha 2]. Retrieved from http://www.iramuteq.org

Segre, C. (1986). Benvenuto Terracini e la linguistica del novecento. In P. Ramat et al. (Eds.), *The history of linguistics in Italy* (pp. 259–276). Amsterdam: John Benjamins.

Severino, A. (1937). *Manuale di nomenclatura linguistica*. Milano: Le lingue estere.

Trevisani, M., & Tuzzi, A. (2015). A portrait of JASA: The history of statistics through analysis of keyword counts in an early scientific journal. *Quality & Quantity, 49*(3), 1287–1304.

Trevisani, M., & Tuzzi, A. (2018). Learning the evolution of disciplines from scientific literature. A functional clustering approach to normalized keyword count trajectories. *Knowledge-Based Systems, 146*, 129–141. https://doi.org/10.1016/j.knosys.2018.01.035.

Chapter 6
The Recent History of Statistics: Comparing Temporal Patterns of Word Clusters

Matilde Trevisani and Arjuna Tuzzi

Contents

6.1 Introduction 106
6.2 Corpus Description and Trends 108
6.3 From Content Mapping to Curve Clustering 109
6.4 Illustration of Cluster Generation and Reading 111
  6.4.1 Pattern 'A': Cluster of Words with Decreasing Trend 113
  6.4.2 Pattern 'C': Cluster of Words of the Classic Era of Statistics 116
  6.4.3 Pattern 'D': Cluster of the Most Popular and Evergreen Words 117
  6.4.4 Pattern 'B': Cluster of Words with Increasing Trend and Emerging 120
6.5 Some Remarks on Normalisation with a Focus on the Cluster of Emerging Words 123
6.6 Discussion and Conclusion 125
Appendix 126
References 129

Abstract The abstracts published by the *Journal of the American Statistical Association* in the time span 1946–2016 have been examined in order to identify relevant timings in the recent history of statistics and retrieve past and current topics that have drawn the attention of one of the most influential communities of statisticians in the world. The focus is on clusters of words that, over time, share a similar trajectory of occurrences in the issues of the journal and on the effect of different choices in the number of clusters. When arrangements in coarser and finer groupings have been compared and contrasted, an interesting nested structure has emerged. Moreover, results have highlighted the conjoint effect of word cycle synchrony and word popularity, which are two of the most important features to be accounted for by the researcher in reading the output of a curve clustering based on observations of word frequencies from a chronological perspective. The research

M. Trevisani (✉)
University of Trieste, Trieste, Italy
e-mail: matilde.trevisani@deams.units.it

A. Tuzzi
Department of Philosophy, Sociology, Education and Applied Psychology,
University of Padova, Padova, Italy

A. Tuzzi (ed.), *Tracing the Life Cycle of Ideas in the Humanities and Social Sciences*, Quantitative Methods in the Humanities and Social Sciences,
https://doi.org/10.1007/978-3-319-97064-6_6

also shows that a knowledge-based system (a computer-based system that supports human learning, endowed with a knowledge-base, a statistical learning engine and a user interface) is able to achieve an effective representation of abstracts and that many elements of the history of statistics may be gleaned by reading the abstracts of a large number of papers and considering 'texts as data'.

Keywords History of statistics · Functional data analysis · Normalisation · Splines · Curve clustering · Cluster number validation

6.1 Introduction

Statistics is a young discipline that initially developed as an instrument to provide governments and public administrators with a reliable picture of the population and its needs. As a response to the administrative and accounting needs of the modern state, economic, political and health statistics have also developed in parallel with demographic and social statistics.

In a previous research study (Trevisani and Tuzzi 2015), we attempted to trace a history of the discipline which, rather than starting from handbooks of the history of statistics,[1] adopted a peculiar perspective. Starting merely from the scientific debate on a mainstream statistical journal, we endeavoured to identify past and current topics covered by statistics, from the perspectives of both methods and application fields. We examined a large set of keywords found in the titles of papers published in the period 1888–2012 by the *Journal of the American Statistical Association* (JASA) and its predecessors.

When we analysed the titles of articles as textual data and looked at the history of the discipline from this viewpoint, we found moments, events and timings that articulate the history of statistics into time spans that are similar to those adopted by Historical Sciences to periodise the history of Europe. Once we decided that the appearance on stage of modern scientific journals represented for statistics the same revolution as the advent of writing in the history of humankind (i.e. it marked the exit from *Prehistory*), we periodised the history of statistics in four phases and considered that:

1. The *Ancient History* of statistics begins at the end of the nineteenth century with the publication of the official journal of the *American Statistical Association* (ASA) in 1888: *Publications of the American Statistical Association* (PASA, 1888–1912). During this period, statistics has a scientific language that is not fully codified and standardised, and its main research topics are wide and diversified. Demography and social statistics are the most explored fields: statisticians collect and examine data with the aim of responding/solving the big

[1] The literature is rich in relevant handbooks used to study the history of Statistics: David and Edwards (2001), Hald (1986, 1998, 2007), Stigler (1986, 1999), Walker (1931), and Westergaard (1932), to cite only a few.

problems of humanity (*wealth, poverty, health, safety, cause of death, infant mortality*) and to study the living conditions and lifestyle of the population. At the end of this first era, PASA gives way to *Quarterly Publications of the American Statistical Association* (QASA, 1912–1921), and in 1922 the *Journal of the American Statistical Association* (JASA) is born.
2. In the 1920s statistics experiences its *Middle Age*, which spans the period of the two World Wars and ends at the end of the 1950s. In this period the economy, the Great Depression and post-war reconstruction dominate the scholars' research interests, and statistics appear to be a true 'political arithmetic'. At the end of this period, rudimentary mathematical instruments appear and give life to a new *Humanism* that is ready to prepare the way for a *Renaissance* of the discipline.
3. In the early 1960s, the *Modern History* begins. Statistics establishes itself as an autonomous discipline thanks to the dissemination of modern statistical tools and develops its own lexicon and specific method. In this period, statistics deals with theories, concepts and methods that are generally considered the essentials of the discipline (*distribution, probability, regression, sampling, test*) and nowadays no longer represent research objects anymore as they are well-established research tools. The development of new methods and techniques runs until the late 1980s, when technological revolutions and the diffusion of modern computers lead statistics to a new era.
4. The last part of the time span represents the *Contemporary History* of statistics, which has conducted statistics since the 1990s until today. In the early 2000s, new problems emerged: high-dimensionality, information complexity and the need for new computational-intensive algorithms and high-performing computers (*dimension reduction, smoothing, functional data analysis, neural networks, mixture model, hierarchical models*), but also interdisciplinarity and hybridisation with other disciplines and new fields of application, especially in the environmental, epidemiological, medical and biomedical fields.

The last two periods of this history, namely *Modern History* and *Contemporary History*, are the most relevant ones in order to study in depth the development of statistical methods 'as we understand them today', and for this reason, we decided to tackle a new study based on the analysis of abstracts published in the last decades. An effective and schematic representation of this storytelling is reported in the Appendix of this chapter.

The abstracts of the articles represent an interesting research object because they express the main contents in few words. Nevertheless, as they do not merely include keywords like titles, they envisage some challenges from the methodological and computational viewpoint.

We have decided to continue to examine the JASA because it still portrays the scientific debate of a large and prestigious community of statisticians in the world today, and it still supports the development of statistics in a broad sense (meetings, publications, education, accreditation, advocacy). Moreover, the JASA still represents one of the world's most relevant premier journals of statistical sciences, and it provides a genuine generalist's perspective and special attention to innovation.

With the appearance of other statistical journal and with the differentiation of statistical fields (economics, demography, epidemiology, etc.), JASA has become less generalistic than before. Nevertheless, it is still considered less specialised than other journals of the field.

6.2 Corpus Description and Trends

Abstracts were not available in the JASA until the 1930s (the only one from the 1930s appeared in 1933) are sporadic in the 1940s and 1950s and gradually become more regular and systematic after the 1960s (Table 6.1), that is, during years that, from this point of view, also mark the beginning of a standardisation of scientific writing.

Table 6.1 The JASA: volumes, issues and available abstracts

Name of the journal	Years	Volumes	Issues	Abstracts	*N*
PASA	1888–1990	1–2	10	0	0
PASA	1891–1895	2–4	16	0	0
PASA	1896–1900	5–7	18	0	0
PASA	1901–1905	7–9	18	0	0
PASA	1906–1910	10–12	20	0	0
PASA	1911–1915	12–14	20	0	0
PASA	1916–1919	15–16	16	0	0
QASA	1920–1921	17	8	0	0
JASA	1922–1925	18–20	16	0	0
JASA	1926–1930	21–25	22	0	0
JASA	1931–1935	26–30	24	1	91
JASA	1936–1940	31–35	21	0	0
JASA	1941–1945	36–40	20	0	0
JASA	1946–1950	41–45	20	143	16,694
JASA	1951–1955	46–50	20	63	7811
JASA	1956–1960	51–55	20	188	24,399
JASA	1961–1965	56–60	20	365	41,769
JASA	1966–1970	61–65	20	556	65,105
JASA	1971–1975	66–70	21	775	62,917
JASA	1976–1980	71–75	20	658	56,277
JASA	1981–1985	76–80	20	570	56,082
JASA	1986–1990	81–85	20	647	134,556
JASA	1991–1995	86–90	20	711	118,861
JASA	1996–2000	91–95	20	636	99,686
JASA	2001–2005	96–100	20	514	92,724
JASA	2006–2010	101–105	20	650	115,351
JASA	2011–2016	106–111	24	745	137,019

Size in word-tokens (*N*)

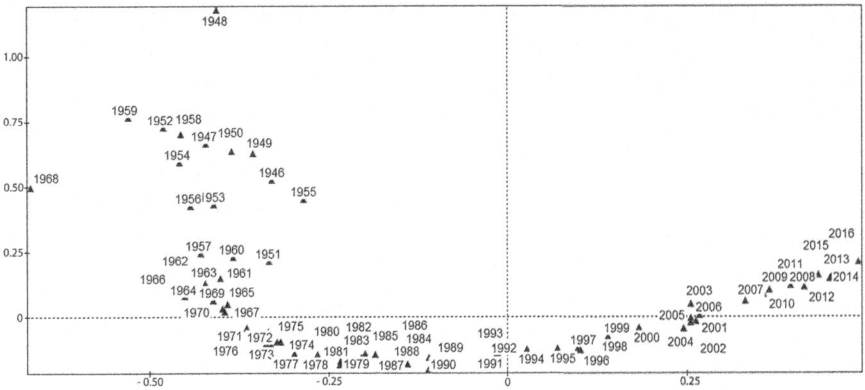

Fig. 6.1 First factorial plane of CA based on 4915 words × 71 years. Projection of years

The abstracts of the papers allow us to observe a more recent period (1946–2016, 71 years) in depth. It represents a little more than half of the time span studied by our previous work that considered only the titles (1888–2012). In the time span of 1946–2016, we retrieved 7221 abstracts. They constitute a large corpus that includes over a million word-tokens ($N = 1{,}029{,}251$) and nearly 27,000 word-types ($V = 26{,}686$). The mean length of an abstract (in terms of word-tokens) is 138 words (min: 10; max: 958; st. dev.: 77 words). For a first explorative data analysis (EDA), we selected words with frequency higher than or equal to 10 and worked with a contingency table of words × years (4915 × 71).

Through the correspondence analysis (CA) of the corpus of abstracts (Fig. 6.1), a new temporal trace unfolds and confirms the existence of a clear temporal pattern in the data. The second axis splits the Cartesian plane into two main half-planes (right and left) around the 1990s, and this division has been already identified in the analysis of titles as a moment of transition from *Modern History* to *Contemporary History*. This analysis, based on abstracts, confirms the previous analysis made on the titles.

In the time span 1888–2012, titles showed a more pronounced reduction in variability over time (Trevisani and Tuzzi 2015), but in more recent times, we can also observe that the scientific language is becoming more and more specialised. The scatter of years on the left side of the graph shows that the range of topics is broader and the lexicon seems richer before the 1960s and 1970s, while the research areas become more limited and the lexicon becomes more technical in recent times, especially in the early years of the twentieth century. Variety is reduced in the name of standardisation, which is also the effect of a process of learning and sharing a 'special language'.

6.3 From Content Mapping to Curve Clustering

In quantitative linguistics, the problem of exploring the temporal evolution of a linguistic phenomenon has often been studied by resorting to linguistic laws (Köhler 2011; Popescu 2009; Tuzzi and Köhler 2015) and time series analysis (Pawlowski et al. 2010). Nevertheless, when trajectories do not portray a regular pattern, these approaches are not able to find a satisfactory solution in terms of goodness of fit and thus achieve results that provide an effective reading of the temporal evolution of word occurrences.

The existence of a latent temporal pattern in word occurrences can be explored by means of CA, which, in our study, reveals a clear time dimension in abstracts and shows that much of the history of statistics may be gleaned by simply reading the abstracts of papers through an EDA. The CA based on the lexical contingency table (words × time-points) is a well-known and established statistical tool in the literature on textual data analysis. When CA is exploited from an exploratory perspective in order to position years and words on a Cartesian plane, it is useful to represent meaningful relationships among words, among time-points and between words and time-points. CA reveals most preeminent timings although it does not highlight how single concepts evolved over time and which words shared the same temporal evolution.

To reconstruct the micro-history of each word and identify words that portray similar temporal patterns, we resort to a functional data analysis (FDA) approach and, within this, to curve clustering (Trevisani and Tuzzi 2015, 2018). A knowledge-based system (KBS) is then proposed to first reconstruct words' life cycles and second, by clustering words with similar life cycles, detect any exemplary temporal patterns representing the latent dynamics of word micro-histories. The major dynamics thus uncovered are then submitted to subject matter experts for interpretation and guidance in the learning process, potentially enabling this to culminate in a conclusive reading (or readings) of the history of the discipline (see Chap. 9 for an extensive description).

In particular, the statistical learning stage of the KBS consists of four steps:

1. Normalising time trajectories of word (raw) frequencies, the type of transformation being chosen according to aspects of life cycles that are considered substantive when comparing words.
2. Filtering time trajectories of word (normalised) frequencies, interpreted as functional data (FD) and thus represented as smooth functions.
3. Curve clustering (CC) to detect all important dynamics underlying the evolution of groups of word micro-histories.
4. Interpretation by expert opinion to decipher detected dynamics and thus compose a narrative of the evolution of the discipline as a whole.

In this study, we have chosen a double normalisation, in particular, d_2 (see Table 9.1, Chap. 9), in order to adjust the uneven document dimension across time (number of texts and their size in word-tokens may vary greatly over time; see Fig. 9.1, Chap. 9) as well as to remedy the great disparity in word popularity (total frequency of individual words in the entire corpus is greatly variable; see Fig. 9.2, Chap. 9), thus enabling the comparison of word curves by timing or synchrony rather than by amplitude. We adopt a basis function approach to filtering with a B-spline basis system (Ramsay and Silverman 2005). Moreover, we take a distance-based approach to CC and use a *k*-means algorithm for FD combined with an appropriate metric for measuring distance between curves (Jacques and Preda 2014; Wang et al. 2016). In this study, we use the Euclidean distance. Lastly, while interpreting, experts can formulate new research questions that may lead to further insights. If CC yields concurrent solutions, the experts can decide on one or more historical narratives for the knowledge field in the period examined.

6.4 Illustration of Cluster Generation and Reading

To achieve a good representation of the trajectories of relevant keywords, we adopted a procedure for the pre-processing of the corpus that envisages an automatic recognition of multiword expressions (see Chap. 8) and then the intersection of the word list with available glossaries for statistical sciences (Trevisani and Tuzzi 2015). Moreover, to reduce the number of items that refer to the same keyword and overcome some of the limitations of an analysis based on word-types, we replaced words with stems. We considered only stems (*estimation*, *model*, *statistics*) and stem segments (e.g. *likelihood estimation*, *mean square error*, *gene expression*) that occur in the corpus of abstracts at least 50 times. The contingency table includes 1351 rows (keywords) and 71 columns (years/volumes). Table 6.2 provides an excerpt of the matrix that reports the occurrences of each keyword in the corpus as a whole and in each time-point. The temporal evolution of a keyword is drawn from the sequence of its occurrences over time, that is, each row of this table portrays a trajectory that from an FDA perspective represents a realisation of an underlying continuous function.

The KBS applied to the corpus produces the best partitions corresponding to the candidates to cluster number that emerged from the pooled validation approach proposed therein (see Fig. 9.7, Chap. 9). In particular, the set of cluster numbers 4, 5, 7 and 15 are subjected to scrutiny (Fig. 6.2).

From the analysis of the partitions selected for the four cluster numbers, it emerges that the more refined groupings are somehow nested in the coarser ones. We then take the four-cluster partition (see top-left panel of Fig. 6.2 and the individual clusters in Fig. 9.8, Chap. 9) as reference as it shows the four fundamental temporal patterns which are gradually more detailed in the finer partitions, namely:

- Pattern 'A': words with decreasing trend from the beginning of the period, generally of low total frequency.
- Pattern 'B': words with increasing trend after 1960 or emerging more recently.
- Pattern 'C': words exhibiting a period of culminant popularity around 1960–1980/1985 and then slowly decreasing.
- Pattern 'D': most popular words everlasting or even with a slowly increasing trend.

In the following paragraphs, each basic cluster will be analysed: first, by examining the nested structure across the four increasingly refined groupings; second, by showing the relevant groups of the finest partition constituting the basic

Table 6.2 Excerpt of the contingency table (stemmed) keywords × years (1351 × 71)

Keywords (stemmed)	Occurrences (corpus)	1946	1947	1948	…	1991	1992	…	2014	2015	2016
estim	7524	5	23	9	:	193	153	:	140	158	137
model	7215	1	1	0	:	129	170	:	202	233	197
data	5107	13	11	10	:	118	139	:	128	156	140
method	4776	19	14	11	:	93	102	:	126	162	134
test	4770	7	16	1	:	82	120	:	44	85	45
distribut	3764	4	9	3	:	59	70	:	34	41	58
propos	3434	0	0	1	:	51	72	:	141	181	153
sampl	3339	19	25	7	:	54	75	:	46	47	41
articl	3109	2	1	8	:	61	79	:	130	151	171
base	2954	8	4	1	:	45	81	:	83	90	77
gener	2692	12	5	3	:	61	82	:	54	41	59
statist	2556	6	9	22	:	48	44	:	45	50	37
studi	2528	3	8	2	:	39	58	:	110	95	82
result	2473	6	14	6	:	35	52	:	43	36	53
procedur	2243	8	5	0	:	40	56	:	45	44	38
function	2169	4	0	1	:	54	35	:	65	48	67
time	2162	7	5	7	:	43	54	:	50	54	51
analysi	2145	6	4	4	:	54	64	:	53	79	61
effect	2089	4	5	5	:	39	40	:	70	65	63
problem	2075	5	8	10	:	47	58	:	29	54	37
regress	2073	0	0	0	:	46	59	:	48	65	41
likelihood estim	602	0	0	0	:	18	11	:	11	9	6
exist	595	1	5	2	:	12	15	:	23	32	18
robust	590	0	0	0	:	19	11	:	8	23	19
limit	590	2	4	4	:	17	11	:	7	18	12
work	588	3	3	2	:	10	16	:	21	16	20
bia	587	4	6	0	:	4	12	:	8	11	5
mean squar error	285	1	0	0	:	8	7	:	1	1	0
statement	50	0	1	0	:	0	1	:	2	0	0
gene express	50	0	0	0	:	0	0	:	1	6	5

cluster; and lastly, by illustrating the typical temporal pattern of these groups through a selection of group words. In establishing the relevance and order of the groups and words presented, we will consider the degree of stability as measured by the multiple Rand index per word and the derived index, on average, per group (see Sect. 9.3.3, Chap. 9). A final summary of the reconstruction of the history of statistics in the period examined is presented in the Appendix of this chapter.

A general overview of the groups in the finest partition, which have been ordered according to the chronological sequence of the four basic patterns (‘A’, ‘C’, ‘D’, ‘B’, that is, from the cluster of words that have tended to disappear to the cluster of

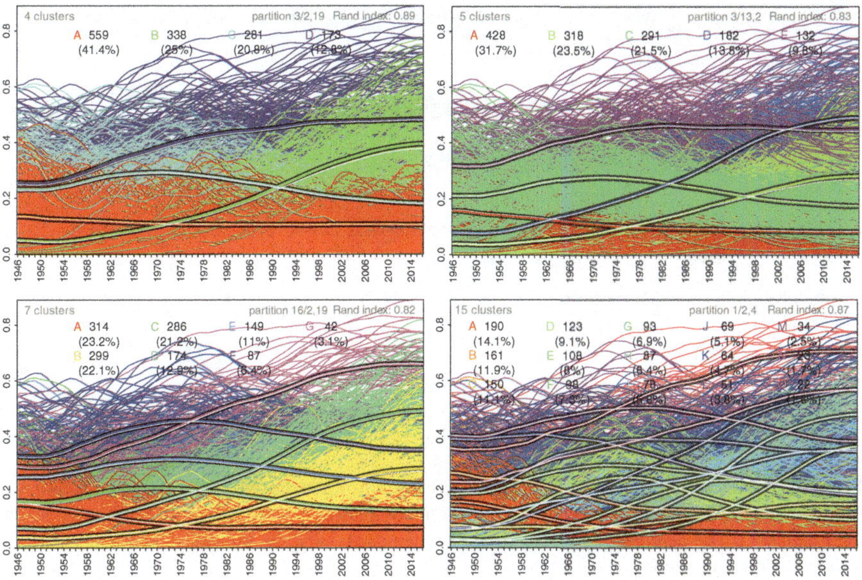

Fig. 6.2 Best partitions corresponding to the set of cluster number candidates in the clustering on d_2 normalised data: the overall 4, 5, 7 and 15 groups

emerging words in the period 1946–2016), is illustrated in Fig. 6.3. For each group, the basic pattern is identified by considering both the direct nesting of the finest partition into the coarsest one and the indirect nesting as derived from the full nesting structure across the four cluster numbers. Whenever a cluster of a finer partition is not entirely nested in a cluster of a coarser partition, the nesting cluster is identified as the one that contains the largest portion of the considered cluster (with a difference from the second largest portion not smaller than 10%, otherwise two nesting clusters are identified). An intermediate pattern may arise wherever the two types of nesting (direct and indirect) give a different basic pattern, as well as wherever two nesting clusters are identified, thus indicating a phase shift (e.g. 'A/C', 'A/B' and 'D/B' in Fig. 6.3).

6.4.1 Pattern 'A': Cluster of Words with Decreasing Trend

Cluster 'A' of the four-group partition contains words that are less and less frequently present, some that eventually disappear, or that were popular quite discontinuously over the period. It produces a 'matrioska'-like structure across the four groupings (Fig. 6.4). The cluster core, which corresponds to A of the 15-group partition (Fig. 6.5), consists mostly of low-frequency words (see panel A in Fig. 6.3) having in general a peak for a short period (right at the beginning, 1946–1950, or during the decade 1950–1960) and a slow decline afterwards, generally disappearing after

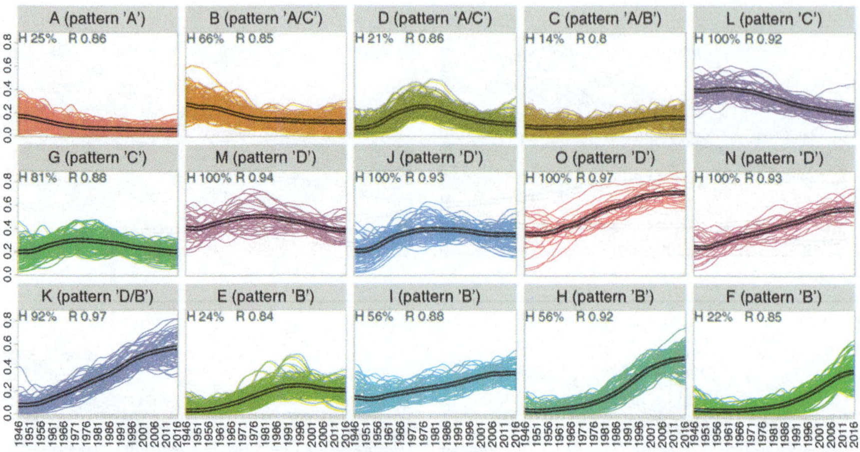

Fig. 6.3 The finest partition into 15 groups ordered according to the chronological sequence of the four basic patterns emerged from the four-cluster partition ('A', 'C', 'D', 'B' and some intermediates). The percentage of high-frequency words (H) and the multiple Rand index (R) are also indicated per group

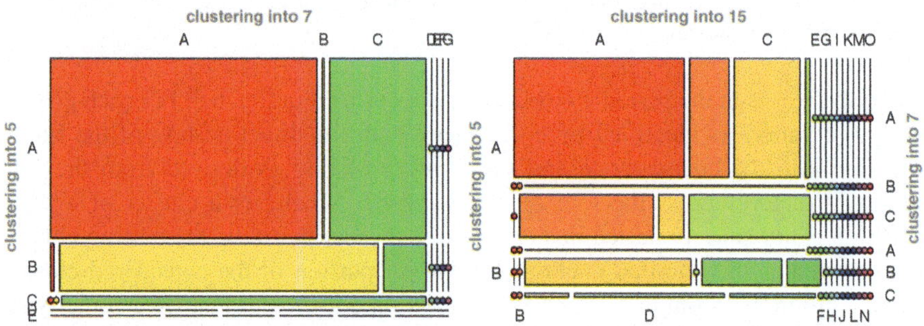

Fig. 6.4 The nested structure of cluster 'A' of the four-group partition: the core A of overall partitions (left and right panels); the rest split in B of both five- and seven-group partitions and C of seven-group partition (left); B, C, D, beyond the core A, of the 15-group partition make up cluster 'A' (right)

about 1975 (Fig. 6.6, label A/A). Constituent words refer to the different declinations of statistics before it became an established discipline (demography: *death*, *mortality*, *fertility*, *insurance*, *demography*; social statistics: *social*, *life*, *household*, *interview*, *school*, *labour*, *migration*, *city*, *women*, *familiar*; institutional statistics: *policy*, *bureau*, *institution*, *country*, *administration*; economic statistics: *employment*, *expenditure*, *manufacture*, *firm*, *earning*, *investment*, *agriculture*, *consumption*, *wage*) and to first tools and technical words of statistics (data collection and design of experiments: *interview*, *universe*, *block*, *stratification*, *sampling design*; descriptive statistics: *quality*, *chart*, *tabular*, *row*, *column*, *cumulative*, *percentage*, *summary*; inferential tools: *method of estimation*, *confidence limit*, *sign test*, *fisher*, *distribution free*, *point estimation*, *probit*, *logit*, *failure rate*, *number of*

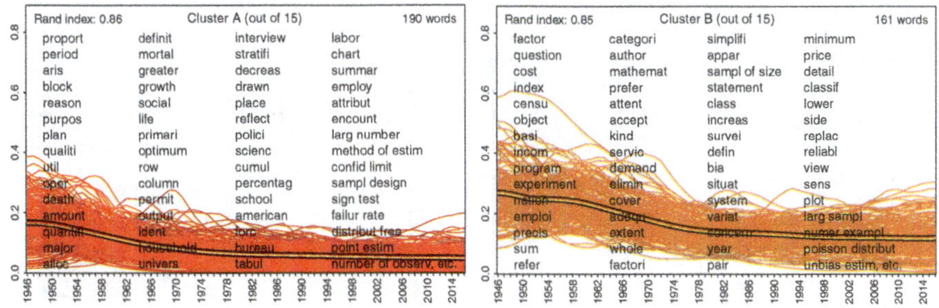

Fig. 6.5 Some clusters of the 15-group partition representing pattern 'A': the core A (left) and the transient (to pattern 'C') B (right)

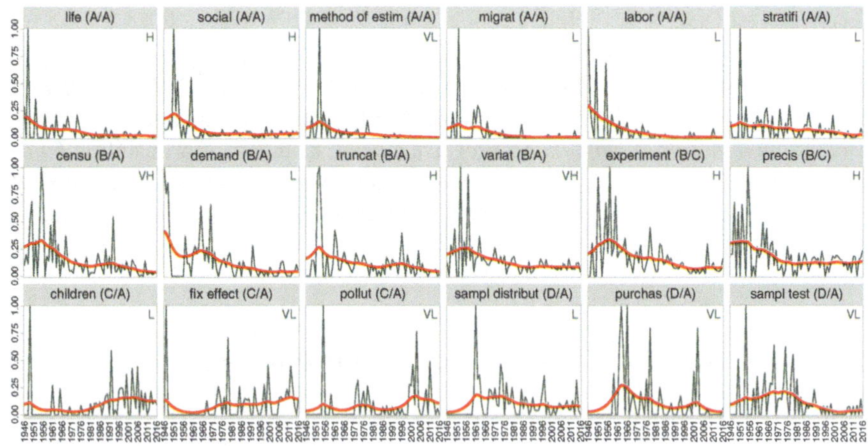

Fig. 6.6 Instances of keywords of cluster 'A' for each nested cluster of the 15-group partition: A, B, C and D. Normalised frequency trajectories and fitted curves. The frequency class is indicated for each keyword: very high (VH), high (H), low (L), very low (VL) (The same reading applies for Figs. 6.9, 6.12, 6.15, and 6.16)

observations; common words: *purpose*, *reason*, *definition*, *utility*, *circumstances*, *judge*, *efforts*, *wish*). The remaining part of 'A' is constituted by clusters B, C and D of the 15-group partition (Fig. 6.4). However, both B (to a minor degree) and D show a progressive shift toward 'C', the next pattern in chronological order, whereas C even recalls pattern 'B' (Fig. 6.3). Note also a lower multiple Rand index for C that indicates to some extent cluster instability. As a side note, we add that C and D are mostly composed of low-frequency words, thus generating very discontinuous trajectories (Fig. 6.6, labels C/A and D/A). Then, we omit a detailed analysis for C, postpone it for D and dwell here briefly on B. It consists of a mix of popular and less popular words which were more frequent during 1946–1970, but that declined afterwards though they have never vanished (Fig. 6.6, label B/A). We can recognise words that refer to economic statistics (*income*, *demand*, *price*, *index*, *cost*) and

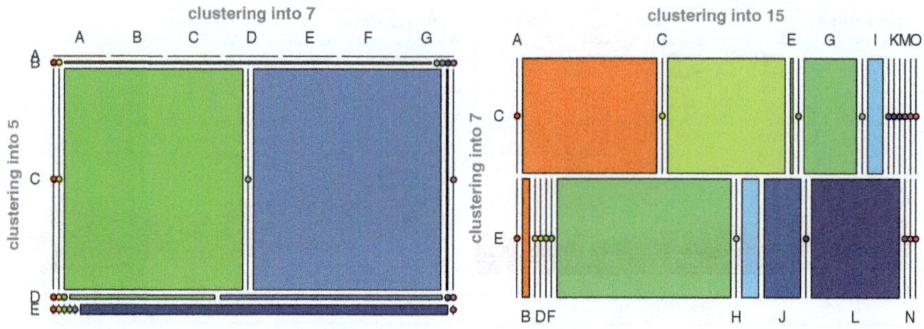

Fig. 6.7 The nested structure of cluster 'C' of the four-group partition: it practically coincides with C of the five-group partition, while it is split into C and E of the seven-group partition (left); couples (B, D) and (G, L) of the 15-group partition make up C and, respectively, E of the seven-group partition (right)

basics of statistics (surveys and sampling: *census*, *survey*, *sampling of size*, *reliable*; design of experiments: *experiment*, *factorial*; principles of estimation: *bias*, *unbiased estimation*, *large sample*, *accuracy*; *categorial*; elements of probability: *binomial*, *poisson distribution*; *truncation*; common words: *statistician*, *numerical example*, *question*, *statement*, *situation*, *concept*, *principle*, *systematic*, *instance*, *agreement*, *support*).

6.4.2 Pattern 'C': Cluster of Words of the Classic Era of Statistics

Cluster 'C' of the four-group partition contains words that had a period of peak popularity around 1960–1980 and since have seen a constant, more or less rapid, decline. As regards its nesting structure, it practically coincides with cluster C of the five-group partition, whereas it is split into C and E of the seven-group partition. Lastly, at the innermost level, couples (B, D) and (G, L) of the 15-group partition are the constituting groups of C and, respectively, E of the seven-group partition (Figs. 6.7 and 6.2). B and D of the 15-group partition are connecting groups between patterns 'A' and 'C' (Fig. 6.3). In particular, D is mostly composed of low-frequency words which exhibit a culminant popularity over 1960–1985 (Fig. 6.9) and refer to theory of estimation and hypopthesis testing (*problem of estimation*, *unknown parameter*, *mean square error*, *minimax*, *ML*, *asymptotic efficiency*, *sample mean*, *trimmed*, *weighted average*, *significance level*, *null distribution*, *Tukey*, *Student*, *Wald*, *Stein*, *goodness-of-fit test*; *Monte Carlo study*—as simulation tools), linear regression (*linear function*, *sum of squares*, *linear estimation*, *linear unbiased estimation*, *dependent variable*, *disturbance*) and other ordinary tools (*contingency*, *matrices*, *order statistics*; probability distributions: *identical/empirical/sampling distribution*, *multivariate normal*, *skew*, *compound*, *gamma*, *beta*, *Bernoulli*, *normal*

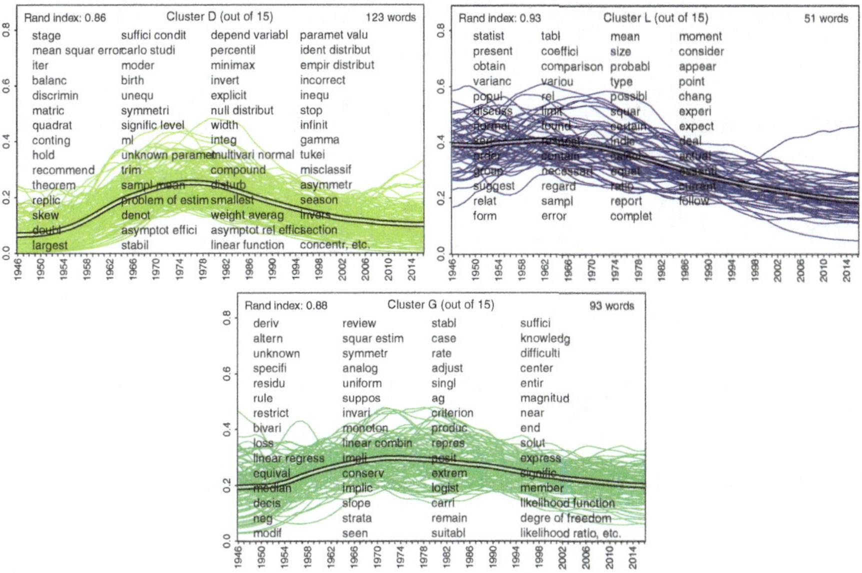

Fig. 6.8 Clusters of 15-group partition representing pattern 'C': D (transient from pattern 'A'), L and G

approximation; time series analysis: *econometrics*, *autocorrelation*, *serial correlation*, *season*, *lag*; common words: *theorem*, *replication*, *hold*, *recommend*) of classic statistics (Fig. 6.8).

A similar temporal pattern is found for L and G, which, however, are almost entirely composed of high-frequency words (Fig. 6.9). These refer to the founder concepts of descriptive statistics and probability (in L, e.g. *statistics*, *population*, *normal*, *series*, *order*, *variance*, *mean*, *sample*, *size*, *error*, *probability*, *moment*; common words: *obtain*, *discuss*, *suggest*, *comparison*, *contain*, *table*) as well as of the bases of inference and linear models (in G, e.g. *linear regression*, *square estimation*, *residual*, *linear combination*, *bivariate*, *univariate median*, *significance*, *sufficiency*, *accuracy*, *efficiency*, *logistic*, *likelihood function*, *likelihood ratio*, *degree of freedom*; common words: *rule*, *restrict*, *loss*, *decision*, *equivalent*, *suppose*, *invariant*, *implication*).

6.4.3 *Pattern 'D': Cluster of the Most Popular and Evergreen Words*

Cluster 'D' of the four-group partition contains only words of very high frequency (panels M, J, O and N in Fig. 6.3), some reaching their peak of popularity over the period 1965–1985, stabilising afterwards to a level only slightly inferior, and others

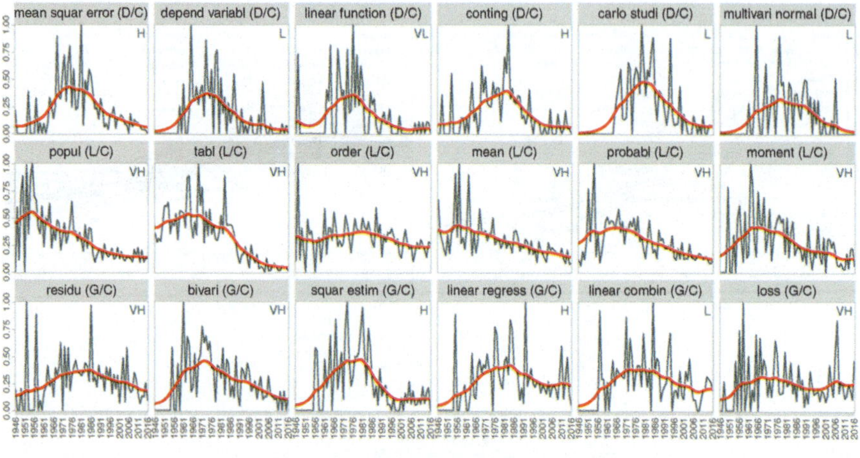

Fig. 6.9 Instances of keywords of cluster ‘C’ for each nested cluster of the 15-group partition: D, L and G

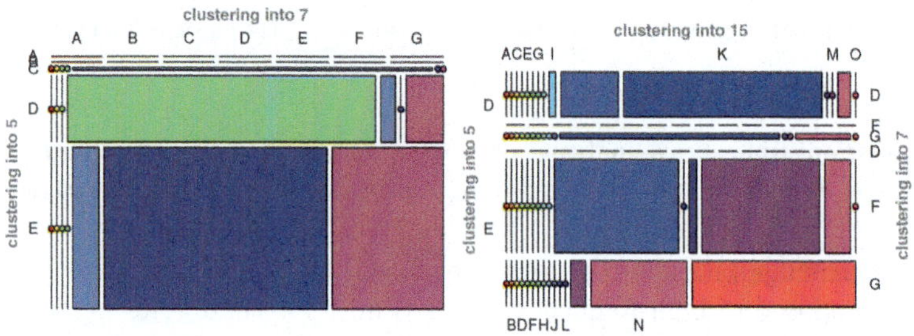

Fig. 6.10 The nested structure of cluster ‘D’ of the four-group partition: it is mostly represented by E of the five-group partition, which is split into F and G of the seven-group partition (left) [Note that the core D of the seven-group partition (left) is a smaller component relatively to the core D of cluster ‘B’ (Fig. 6.13)]; couples (J, M) and (N, O) of the 15-group partition make up F and, respectively, G of the seven-group partition (right)

being in constant increase, thus replacing the classic lexicon of statistics in the contemporary age.

As regards its nesting structure, it is mostly represented by cluster E of the five-group partition, which, in turn, is split into F and G of the seven-group partition; lastly, at the innermost level, couples (J, M) and (N, O) of the 15-group partition are the constituting groups of F and, respectively, G of the seven-group partition (Figs. 6.10 and 6.2). Note that cluster K of the 15-group partition is a concatenating group between patterns ‘D’ and ‘B’, as will be illustrated in the next paragraph (Figs. 6.13–6.15).

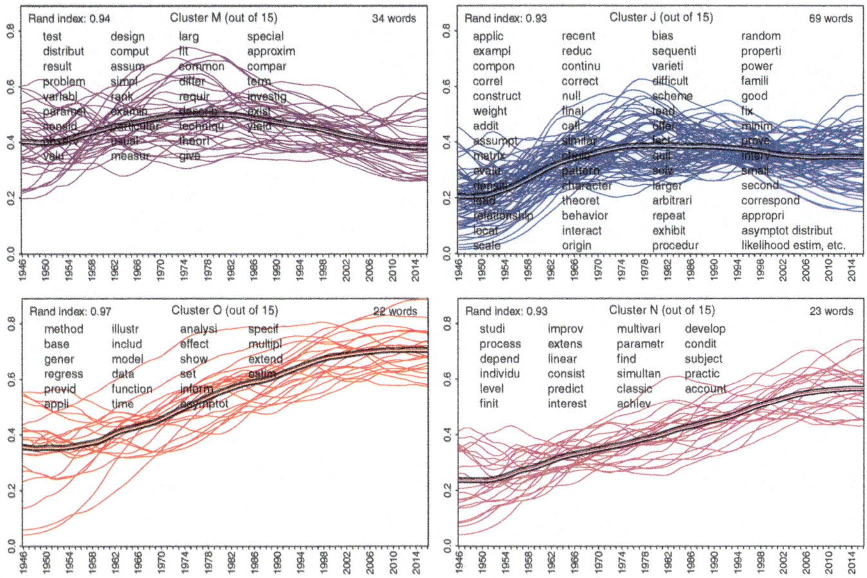

Fig. 6.11 Clusters of 15-group partition representing pattern 'D': M, J, O and N

M contains basic words which were dominant over 1965–1985, stabilising after that time at a lower level or currently decreasing, such as founding terms of traditional lexicon (*test*, *distribution*, *variable*, *parameter*, *design*, *rank*, *observation*, *fit*, *measure*) and common terms (*problem*, *result*, *assume*, *compute*, *examine*, *theory*, *technique*, *require*, *differ*, *approximation*, *compare*, *investigation*); J features words which peaked at around 1975/80 but have not lost their vitality over time (*correlation*, *matrix*, *weight*, *relationship*, *similar*, *density*, *continuous*, *location*, *scale*, *choice*, *pattern*, *character*, *interaction*, *bias*, *correct*, *additive*, *repeat*, *correspondence*, *independence*, *asymptotic distribution*, *likelihood estimation*; as for the language: *application*, *example*, *component*, *construct*, *assumption*, *evaluate*, *lead*, *reduce*, *theoretical*, *pattern*, *behaviour*, *scheme*, *tend*, *offer*, *arbitrary*, *exhibit*, *procedure*, *prove*); O represents fundamental words continuously increasing up to 2000 and then stabilising (*model*, *data*, *regression*, *function*, *asymptotic*, *multiple*, *effect*, *estimation*; as for the language: *method*, *analysis*, *apply*, *illustration*, *provide*, *inform*, *extend*); lastly, N includes words that, although classical, have not lost vital force and are still unattainable when composing a statistical text (*process*, *dependent*, *individual*, *level*, *linear*, *parametric*, *consistent*, *predict*, *multivariate*, *simultaneous*, *subject*; as for the language: *study*, *improvement*, *extension*, *interest*, *find*, *achieve*, *develop*, *practice*, *account*) (see Figs. 6.11 and 6.12).

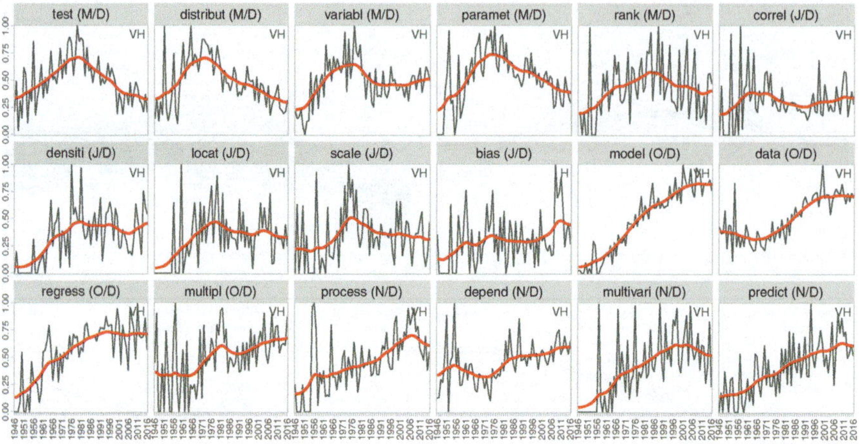

Fig. 6.12 Instances of keywords of cluster 'D' for each nested cluster of the 15-group partition: M, J, O and N

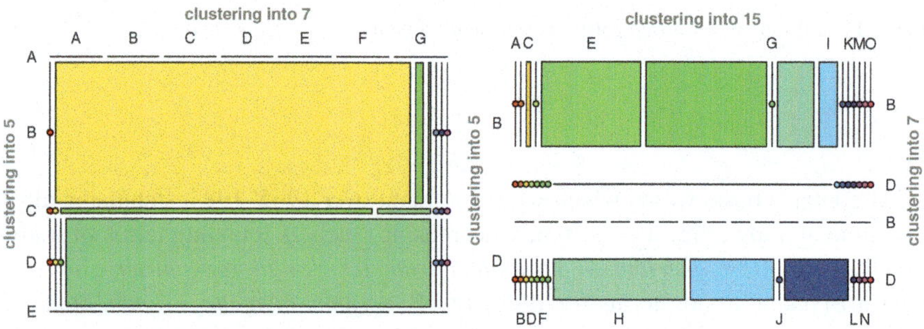

Fig. 6.13 The nested structure of cluster 'B' of the four-group partition: it is split into (B, D) of both the five-group and seven-group partitions (left); (E, F) and (H, I, K) of the 15-group partition make up B and D, respectively, of the seven-group partition (right)

6.4.4 *Pattern 'B': Cluster of Words with Increasing Trend and Emerging*

Cluster 'B' of the four-group partition contains the contemporary bag of words of statistics: words that have increased their popularity especially after 1990, that have already begun the descending parable after 2000 or that are emerging in the last 10 years. As regards its nesting structure, it is split into B and D of both the five-group and seven-group partitions. Then, B and D, which almost match in these last two groupings, are roughly split into (E, F) and, respectively, (H, I, K) of the 15-group partition (Fig. 6.13).

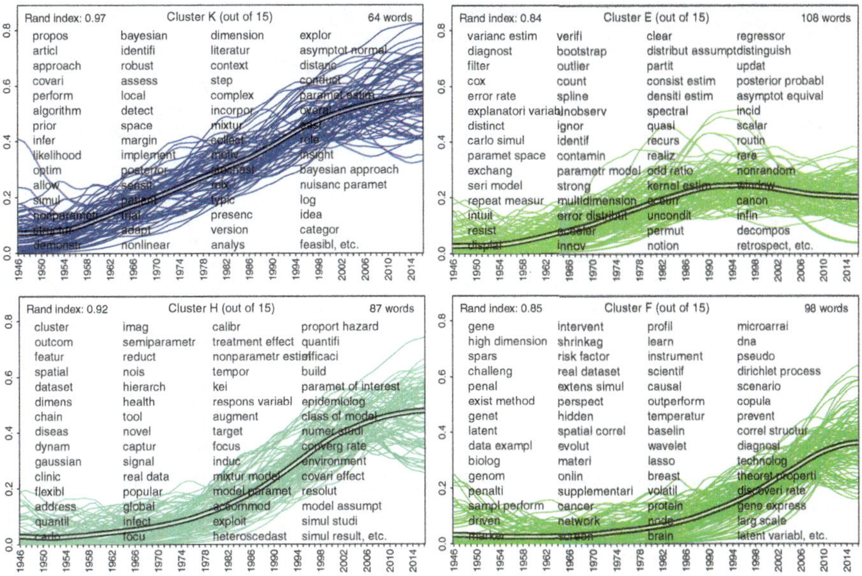

Fig. 6.14 Clusters of 15-group partition representing patterns 'B': K (transient from 'D'), E, H and F

Clusters K and H consist mostly of high frequency words with a continuous markedly increasing trend of popularity (Fig. 6.3). K includes very high-frequency words that suggest the evolution of approaches (e.g. *likelihood*, *bayesian*, *nonparametric*, *robust*, *local*, *adaptive*, *nonlinear*, *sensitivity* analysis, *distance-*, *algorithm-* and *simulation*-based), of modelling problems (e.g. *dimension*, *complex*, *space*, *mixture*, *mixed*, *survival*) and of language (e.g. *propose*, *perform*, *assess*, *detect*, *implement*, *explore*) since 1960, while H features words that gained popularity generally at the end of 1990 and translate those ideas into new models (*flexible*, *hierarchical*, *heteroscedastic*, *dynamic*, *temporal*, *spatial*, *capture*, *proportional hazard* and *mixture model*), methods (Monte *Carlo*, *augmentation*, *signal* processing, *smoothing*, *nonparametric*, *semiparametric*, *kernel*, *reduction*, *clustering*, *matching*, *calibration*, *image* and *longitudinal* data, *tree* models) and studies (*health* statistics, *clinic* trials, *infectious*, *disease*, *environmental* statistics, *dose*, *exposure*, *epidemiology*) (Fig. 6.14). Note that K is a concatenation group between patterns 'D' and 'B' being a cluster of words born with the consolidation of statistics that have acquired a high level of popularity over time (pattern 'D') but also constituting the specialised terms for dealing with the new themes and applications of the contemporary age (pattern 'B'). We omit a detailed description of cluster I, which consists of words having a more fluctuating popularity although stabilised around 2000, as it features less substantial words with respect to K and H.

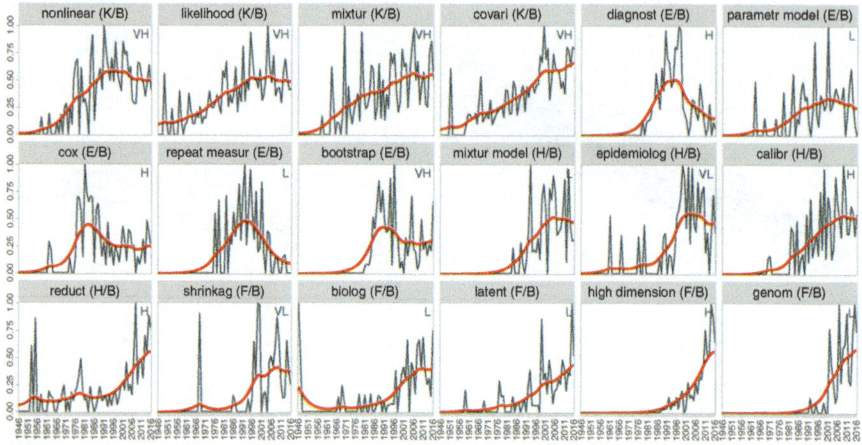

Fig. 6.15 Instances of keywords of cluster 'B' for each nested cluster of the 15-group partition: K, E, H and F

Clusters E and F consist mostly of low-frequency words with an increasing trend. Those contained in E suggest themes (*bootstrap*, *jacknife*, Monte C*arlo simulation* and *method*, *density* and *kernel estimation*, *variance estimation*, *consistent estimation*, *filtering*, *splines*, *smoother*, *outlier* detection, *diagnostic* tools, *missing values*, *nonrandom*, *quasi*-likelihood, *contamination* methods, *parametric* models, *error distribution*, *distributional assumption*, *asymptotic equivalence*, *multidimensional* analysis, *count* data, *odds ratio*, *spectral* analysis, time *series model*, *repeated measures*, *exchangeable* model, *cox*, *hazard* function, *retrospective* studies) that gained popularity until the late 1990s, after which they began to decline, while those in F represent issues (*high dimension*, *sparsity*, *latent* and *hidden* process, *heterogeneity*, *correlation structure*, *volatility*, *spatial correlation*, *risk factors*, *mapping*, *microarray*), approaches (nonparametric, e.g. *wavelets*; simulation-based, e.g. *sampler*, *slice* sampling, *extensive simulation*; machine *learning*, *data-driven*), estimation methods (*shrinkage*, *penalisation*, *penalty*, *instrumental variables*, *regularisation*, *sparse estimation*, *lasso*), models (e.g. *latent variable*, *causal*, *semiparametric model*, *copula*, *trait*, *trajectory*, *dirichlet process*, graphical models and *network* analysis), research areas (medicine, e.g. *virus*; public health, e.g. *prevention*, *intervention*; environmetrics, e.g. *environment*, *temperature*; *biology*; epidemiology, e.g. *prevalence*; biomedicine, e.g. *cancer*, *breast*, *brain*, *gene*, *dna*, *genetics*, *marker*, *genoma*, *protein*, *gene expression*) and 'common saying' (*challenge*, *goal*, *task*, *profile*, *perspective*, *scenario*, *realistic*) that have a surge after 2000 since they have a rebirth in recent times or are emerging (Figs. 6.14 and 6.15).

6.5 Some Remarks on Normalisation with a Focus on the Cluster of Emerging Words

For a better understanding of clustering results, let us examine some effects of the word frequency normalisation adopted in this study. We recall that a transformation of raw frequencies of words is necessary to correctly reconstruct and compare the temporal evolution of words. A form of normalisation by time-point should be regarded as preliminary in order to adjust the uneven size of subcorpora across time (see Fig. 9.1, Chap. 9). A further form of normalisation by word might be appropriate in order to regulate the great disparity in word popularity, which produces very strong asymmetry of frequency spectrum by time-point and sparseness of low-frequency word trajectories (see Fig. 9.2, Chap. 9).

In this study, we have chosen a double normalisation (d_2, see Table 9.1, Chap. 9) that normalises both by time-point and by word, in particular, from dividing the raw frequency of a word at each time-point/volume, both by the total number of word-tokens in each volume and by the maximum frequency of the word trajectory. In particular, this last normalisation (by word) is able to substantially reduce the high skewness featuring the bundle of word trajectories. However, it cannot completely remedy the problem of sparsity. As a result, a trace of word popularity remains and continues to influence the comparison between trajectories as we describe below.

The criterion chosen in the illustration of temporal patterns (both in subsection sequence of Sect. 6.4 and in Fig. 6.3, namely ‘A’, ‘C’, ‘D’, ‘B’) is the chronological one: clusters are ordered according to the exemplary life cycle of group words from the one that has already been concluded (‘A’) to the one that has recently begun (‘B’). As well, clusters of the 15-group partition are chronologically ordered within each of the four basic temporal patterns (Fig. 6.3). However, a chronological reading is not sufficient to discriminate the four patterns and the more analytical patterns of the 15-group partition that compose them. A second key to reading is the level of popularity featuring the words of the considered cluster. In fact, note how some patterns have a relatively parallel gait and are mostly distinguished by the height of the curves (see B-L, D-G, C-I, F-H-K in Fig. 6.3). This result is due to the effect that the chosen normalisation has on the filtering of curves and therefore on their grouping on the basis of similarity. Namely, words with a low or very low total frequency tend to have sparse trajectories, i.e. to have zero or almost zero frequency for relatively long stretches of the period, either continuous (in the case that the word concludes or begins its life cycle; see, e.g. *semiparametric model* in Fig. 6.16, third row, left-most panel) or intermittent (giving rise to peak-and-valley trajectories; see, e.g. *realistic*, sixth row, left-most panel), and very high differences of height along the trajectory (being frequency values little spaced). This involves that the smoothing of the trajectory tends to produce a flattened curve downwards. On the contrary, words with a high or very high total frequency tend to have non-negligible frequencies for most of the period and trajectories with lower differences of height (being the grid of frequency values finer) (see, e.g. *nonparametric* and *simulation*, left-most panels of first row).

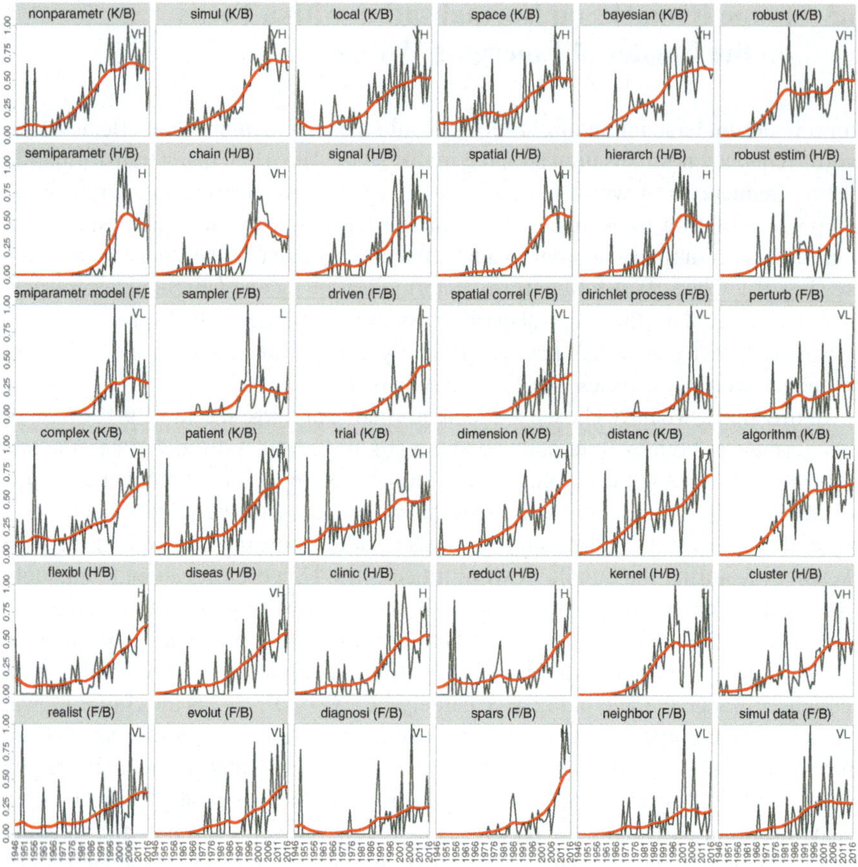

Fig. 6.16 Instances of words of cluster 'B' from clusters K, H and F of the 15-group partition; 12 themes illustrated by 12 (vertical) trios of words taken, in order, from K, H and F. For example, the first top-left trio *nonparametric* (K), *semiparametric* (H), *semiparameric model* (F) refers to theme 'nonparametric methods'

Let us illustrate the point considering pattern 'B', that is, the cluster of words that constitute the contemporary lexicon of statistics and that evoke themes emerging in more recent times. In particular, clusters K, H and F of the 15-group partition (Fig. 6.14) feature a relatively synchronised pattern, which is essentially distinguished by a different height. They comprise, respectively, 8%, 44% and 78% low-frequency words (Fig. 6.3). Figure 6.16 shows a sample of words which refer to 12 themes, each exemplified by a trio of words taken, in order, from clusters K, H and F, to compare the effect of popularity level on curve smoothing. The themes are nonparametric methods, simulation-based estimation, data-denoising and data-driven approach, spatial models, Bayesian models, robust estimation, complex and flexible models, epidemiology, statistical medicine, high-dimension and sparsity, local estimation and machine learning. We can take any of the 12 trios (K, H, F) to see how the word popularity affects the height of smoothed curve, the shape being similar. For example, compare sparsity and bumpiness of trajectories for *space* (K), *spatial* (H) and *spatial correlation* (F) keywords (the fourth vertical trio in the first row, Fig. 6.15), which are, the first two, very-high (VH) and, the third, very-low (VL) frequency words: the smoothed curves are somehow synchronised although they are different in level.

6.6 Discussion and Conclusion

In a previous study, we observed the existence of a clear temporal pattern in articles of JASA, and we showed that a large share of relevant elements can be retrieved through the statistical analyses of the titles of papers published by JASA (Trevisani and Tuzzi 2015). In this new study, we had the opportunity to elaborate on more recent history by a (distant) reading of the abstracts. Through CA results, a clear temporal pattern emerges (Fig. 6.1), even if, beyond this pattern, it is difficult to understand how individual concepts evolved over time and, above all, which keywords share the same temporal development. In brief, CA proves useful to obtain a general overview of the main contents of the corpus, but curve recognition and clustering are necessary to trace the individual life cycles of keywords and find common dynamics latent to word micro-histories.

The proposed KBS leads to the identification of a number of possible partitions of the corpus in word clusters (Fig. 6.2). Our procedure for cluster number selection, in fact, produces a set of candidates to cluster number: in this study, 4, 5, 7 and 15. From an analysis of the agreement between the concurrent partitions (Wagner and Wagner 2007), it emerges that, for this corpus, the finer partitions are essentially nested in the coarser ones. From this finding, the reading of results is based on the four basic temporal patterns of the coarsest partition, each of which is analysed in depth by the examination of the nested clusters in the 15-group partition.

The reading follows the chronological order, that is, from the cluster of words that have tended to disappear to the cluster of emerging words in the period 1946–2016, both in the analysis of the sequence of the four fundamental patterns and in that of the sequence of the 15 nested sub-patterns (Fig. 6.3).

However, the chronological reading key must be integrated with the information on the popularity level of cluster words. In fact, on the one hand, the adopted double normalisation substantially solves the problem of strong asymmetry of the frequency distribution due to the enormous difference between popular and rare words, but on the other hand, it does not remedy the problem of sparseness in the trajectories of unpopular words. Therefore, a trace of the cluster words' popularity remains in the reconstruction of the temporal pattern. This reflects, on the one hand, the synchrony of curves, and on the other hand, the popularity level of cluster words. An example of this effect of the normalisation is offered by the case of the almost parallel temporal patterns, but of different heights, that feature the clusters of words evoking emerging themes (see Sect. 6.5).

Appendix

Reconstruction of the history of statistics in the period 1946–2016

Middle age	Until about 1960	Pattern	Group	Chronology	Popularity	Themes	Instances of keywords
		'A'	A	Peak during 1946–1950 or 1950–1960, slow decline afterwards, generally disappearing after about 1975	VH/H 3/22% L/**VL** 33/42%	Demography, social/institutional/economic statistics; data collection and design of experiments; first tools of descriptive and inferential statistics; (common words)	*death, mortality, fertility, insurance, demography; social, life, household, interview, school, labour, migration, city, women, familiar; policy, bureau, institution, country, administration; employment, expenditure, manufacture, firm, earning, investment, agriculture, consumption, wage; interview, universe, block, stratification, sampling design; quality, chart, tabular, row, column, cumulative, percentage, summary; method of estimation, confidence limit, sign test, fisher, distribution free, point estimation; probit, logit, failure rate; (purpose, reason, definition, utility, circumstances, judge, efforts, wish)*
		'A/C'	B	Peak during 1946–1970, decline afterwards though never vanishing	VH/**H** 23/43% L/VL 24/10%	Economic statistics; surveys and sampling, design of experiments, elements of probability, principles of estimation	*income, demand, price, index, cost; census, survey, sampling of size, reliable; experiment, factorial; categorial; binomial, poisson distribution, truncation; precision, bias, unbiased estimation, large sample, accuracy; (statistician, numerical example, question, statement, situation, concept, principle, systematic, <u>instance</u>, agreement, support)*
Modern history	1960–1990		D	Dominant over 1960–85 since then rapid decline	VH/H 1/20% **L**/VL 44/35%	Classic theory of estimation and hypothesis testing, linear regression, probability distributions, descriptive tools, time series analysis	*problem of estimation, unknown parameter, mean square error, minimax, ML, asymptotic efficiency, sample mean, trimmed, weighted average, significance level, null distribution, Tukey, Student, Wald, Stein, goodness-of-fit test, population mean; linear function, sum of squares, linear estimation, linear unbiased estimation, dependent variable, disturbance; Monte Carlo study; identical distribution, empirical distribution, sampling distribution, multivariate normal, skew, compound, gamma, beta, Bernoulli, normal approximation; contingency, matrices, order statistics; econometrics, autocorrelation, serial correlation, season, lag; (theorem, replication, hold, recommend)*

		'C'	L	Dominant up to 1970 since then slow decline	**VH**/H 84/16%	Founding concepts of descriptive statistics and probability	*statistics, population, normal, series, order, variance, mean, sample, size, error, probability, moment; (obtain, discuss, suggest, comparison, contain, table)*
			G	Dominant over 1960–1980/1990 since then slow decline	VH/**H** 35/45% L/VL 18/1%	Basics of inference and linear models	*linear regression, square estimation, residual, linear combination, bivariate, univariate median, significance, sufficiency, accuracy, efficiency, logistic, likelihood function, likelihood ratio, degree of freedom; (rule, restrict, loss, decision, equivalent, suppose, invariant, implication)*
		'D'	M	Dominant over 1965–1985 since then slow decline or stabilised at a lower level	**VH** 100%	Founding concepts of classic statistics	*test, distribution, variable, parameter, design, rank, observation, measure, fit; (problem, result, assume, compute, examine, theory, technique, require, differ, approximation, compare, investigation)*
			J	Peak at ~1975/80, stabilised afterwards	**VH**/H 68/32%	Toolbox of statistical analysis up to contemporary age	*correlation, matrix, weight, relationship, similar, density, continuous, location, scale, choice, character, interaction, bias, correct, additive, repeat, correspondence, independence, asymptotic distribution, likelihood estimation; (application, example, component, construct, assumption, evaluate, lead, reduce, theoretical, pattern, behaviour, scheme, tend, offer, arbitrary, exhibit, procedure, prove)*
Contemporary history	1990-nowadays		O	Steady increase up to 2000, stabilised afterwards	**VH** 100%	Fundamental words of statistics even more in the contemporary age	*model, data, regression, function, asymptotic, multiple, effect, estimation; (method, analysis, apply, illustration, provide, inform, extend)*
			N	Steady increase up to 2005/10, stabilised afterwards	**VH**/H 96/4%	Classical although unattainable words in a statistical text	*process, dependent, individual, level, linear, parametric, consistent, predict, multivariate, simultaneous, subject; (study, improvement, extension, interest, find, achieve, develop, practice, account)*

		'D/B'	K	Marked increase since 1960	**VH**/H 70/22% L 8%	Approaches and modelling problems of contemporary statistics	*inference, likelihood, bayesian, prior, posterior, bayesian approach, nonparametric, robust, local, adaptive, nonlinear, distance, algorithm, simulation, asymptotic normality, sensitivity; dimension, complex, space, mixture, mixed, survival, patient, trial*; (*propose, approach, perform, assess, detect, implement, analyse, explore*)
Contemporary history	1990-nowadays	'B'	H	Marked increase since after 1990	VH/**H** 13/43% L/VL 23/21%	Models and methods of contemporary statistics, medical/health/ environmental statistics, epidemiology	*flexible, hierarchical, heteroscedastic, dynamic, temporal, spatial, capture, proportional hazard, mixture model*, Monte *Carlo, chain, augmentation, signal, smoothing, nonparametric, semiparametric, kernel, reduction, clustering, matching, calibration, image, longitudinal* data, *trees*; *health, clinic* trials, *infectious, disease, environmental, dose, exposure, epidemiology*; (*address, novel, focus, target, accommodate, exploit, build, enhance, highlight, suffer, monitor, real data, package, software*)
			E	Increase from late 1980 until late 1990, decline afterwards	VH/H 4/18% L/**VL** 35/43%	Research areas and models losing vitality in the new millennium	*bootstrap, jacknife*, Monte *Carlo simulation* and *method, density* and *kernel estimation, variance estimation, consistent estimation, filtering, splines, smoother, outliers, diagnostics, missing values, nonrandom, quasi*-likelihood, *contamination* methods, *parametric model, error distribution, distributional assumption, asymptotic equivalence, multidimensional* analysis, *count* data, *odds ratio, spectral* analysis, *time series model, repeated measures, exchangeable* model, *cox, hazard* function, *retrospective* studies; (*unobserved, ignorance, occurrence, realisation, unconditional, notion, display, verify, identify, distinguish*)
			F	Surge after 2000 as a rebirth or since emerging	VH/H 4/20% L/**VL** 32/44%	Issues, approaches, estimation methods, models, research areas of the new millennium, machine learning, medical/health/ environmental statistics, epidemiology, biostatistics	*high dimension, sparsity, latent, hidden* process, *heterogeneity, correlation structure, volatility, spatial correlation, risk factors, mapping, microarray; wavelets, sampler, slice, extensive simulation, learning*, data-*driven*; *shrinkage, penalisation, penalty, instrumental variables, regularisation, sparse estimation, lasso*; *latent variable, causal, semiparametric model, copula, trait, trajectory, dirichlet process, network, segmentation*; *virus, prevention, intervention, environment, temperature, biology, prevalence, cancer, breast, brain, gene, dna, genetics, marker, genoma, protein, gene expression*; (*challenge, goal, task, profile, perspective, scenario, framework, realistic*)

References

David, H. A., & Edwards, A. W. F. (2001). *Annotated readings in the history of statistics*. New York: Springer-Verlag.

Hald, A. (1986). *A history of probability and statistics and their applications before 1750*. New York: Wiley.

Hald, A. (1998). *A history of mathematical statistics from 1750 to 1930*. New York: Wiley.

Hald, A. (2007). *A history of parametric statistical inference from Bernoulli to Fisher, 1713–1935*. New York: Springer.

Jacques, J., & Preda, C. (2014). Functional data clustering: A survey. *Advances in Data Analysis and Classification, 8*(3), 231–255.

Köhler, R. (2011). Laws of languages. In P. C. Hogan (Ed.), *The Cambridge encyclopedia of the language science* (pp. 424–426). Cambridge: Cambridge University Press.

Pawlowski, A., Krajewski, M., & Eder, M. (2010). Time series modelling in the analysis of homeric verse. *Eos, 97*(2), 79–100.

Popescu, I. I. (2009). *Word frequency studies*. Berlin: Mouton De Gruyter.

Ramsay, J., & Silverman, B. W. (2005). *Functional data analysis (Springer series in statistics)*. New York: Springer.

Stigler, S. M. (1986). *The history of statistics. The measurement of uncertainty before 1900*. Cambridge: The Belknap Press of Harvard University Press.

Stigler, S. M. (1999). *Statistics on the table: The History of statistical concepts and methods*. Cambridge: Harvard University Press.

Trevisani, M., & Tuzzi, A. (2015). A portrait of JASA: The History of Statistics through analysis of keyword counts in an early scientific journal. *Quality and Quantity, 49*(3), 1287–1304.

Trevisani, M., & Tuzzi, A. (2018). Learning the evolution of disciplines from scientific literature. A functional clustering approach to normalized keyword count trajectories. *Knowledge-Based Systems, 146*, 129–141.

Tuzzi, A., & Köhler, R. (2015). Tracing the history of words. In A. Tuzzi, M. Benesová, & J. Macutek (Eds.), *Recent contributions to quantitative linguistics* (pp. 203–214). Berlin: DeGruyter.

Wagner, S., & Wagner, D. (2007). Comparing clusterings: An overview. Universitat Karlsruhe, Fakultat fur Informatik Karlsruhe. Retrieved from https://publikationen.bibliothek.kit.edu/1000011477/812079

Walker, H. M. (1931). *Studies in the history of statistical method with special reference to certain educational problems*. Baltimore: The Williams and Wilkins.

Wang, J. L., Chiou, J. M., & Mueller, H. G. (2016). Functional data analysis. *Annual Review of Statistics and Its Application, 3*(1), 257–295.

Westergaard, H. (1932). *Contributions to the history of statistics*. London: P. S. King and son.

Part II
Concepts and Methods

Chapter 7
Treat Texts as Data but Remember They Are Made of Words: Compiling and Pre-processing Corpora

Stefano Ondelli

Contents

7.1 What Is a Corpus? A Preliminary Definition 134
7.2 What's in a Corpus? Corpus Design and Sociolinguistics 135
7.2.1 Diachronic Variation 137
7.2.2 Diatopic Variation 138
7.2.3 Diastratic Variation 139
7.2.4 Diamesic Variation 141
7.2.5 Diaphasic Variation 142
7.3 Preparing a Corpus: Pre-processing 143
7.3.1 Tokenisation 144
7.3.2 Punctuation 145
7.3.3 POS Tagging and Lemmatisation 146
7.3.4 Stemming 148
7.4 Conclusion 148
References 149

Abstract When analysing corpora with automatic and statistical means, one should remember that the raw material being treated is language and the specific nature thereof ought to be considered in all stages of research. Since language cannot be investigated per se, corpora can only reveal the characteristics of limited instances of linguistic behaviour: even exhaustive corpora only supply a finite set of texts which should be assessed in the light of a number of extra-linguistic factors impacting linguistic traits from different viewpoints: the sender's and recipient's region of origin, social and educational background and gender; the channel of communication; the topic under discussion and the formality of the situation, not to speak of the period in history when texts were produced. Such factors come into play in defining the linguistic properties of each single text (fragment) in the corpus, and their overall balance should be considered during the preliminary stages of corpus design and compilation.

S. Ondelli (✉)
University of Trieste, Trieste, Italy
e-mail: sondelli@units.it

A. Tuzzi (ed.), *Tracing the Life Cycle of Ideas in the Humanities and Social Sciences*, Quantitative Methods in the Humanities and Social Sciences,
https://doi.org/10.1007/978-3-319-97064-6_7

After having made decisions in terms of the selection of the texts to be included in the corpus, linguistic data need to be prepared for automatic processing. This stage too is far from intuitive and automatic: from the very identification of tokens of language to the extraction of lemmas, researchers should take into account qualitative aspects. Both corpus compilation and pre-processing cannot be considered neutral operations with a view to the results of automatic analysis and should be made explicit to enable the assessment of results and further exploitation of the same corpus.

Keywords Corpus linguistics · Sociolinguistics · Language variation · Corpus design · Corpus pre-processing

7.1 What Is a Corpus? A Preliminary Definition

Over the last 20 years or so, corpus linguistics has become so popular (or notorious) that even researchers not versed in this field of study often refer to corpora. Statements such as "I have collected a corpus of essays written by students (or newspaper articles, or love poems etc.)" are heard very commonly without any further reference to how those texts have been selected and processed for research purposes. According to this approach, a loose definition of "corpus" is "a collection of texts having something in common". This common feature may be their author (Walter Scott's novels, but also texts written by men vs. texts written by women), genre (newspaper articles, poems), communication channel (oral vs. written), format (electronic vs. paper), topic (scientific vs. legal texts), etc. However, this loose definition seems to fail to account for the novelty of the discipline called "corpus linguistics": after all, a great deal of research on oral or written communication has been based on the analysis of a number of texts (pre-electronic corpora). For example, the word list and definitions of the *Vocabolario dell'Accademia della Crusca* (1612; http://vocabolario.sns.it/html/index.html) were extracted from a selected corpus of literary authors. However, we can probably start to talk about "corpus linguistics" proper in the 1960s, with the growing availability of texts in electronic format, as witnessed by the publication and first analyses of the *Brown University Standard Corpus of Present-Day American English* (http://www.nltk.org/nltk_data/).

Among the many available in the relevant literature, a detailed definition of "corpus" was provided by Manuel Barbera et al. (2007, p. 70; my translation):

> A corpus is a finite collection of (written, oral or multimodal) texts or parts thereof in electronic format, consistently processed (i.e. tokenised and added with adequate mark-up) so as to be treated and investigated automatically by means of software. If (as is often the case) analyses are conducted for linguistic purposes (e.g. the description of natural languages or their registers), texts are generally selected according to their authenticity and representativeness.

This can be considered a "hard" definition of a corpus since it implies specific procedures to identify units of language (i.e. tokenisation, see Sect. 7.3.1) enriched with additional information (mark-up). These procedures can be carried out automatically thanks to dedicated software (but the price to pay is a significant error rate) or manually, in which case they are very time-consuming and one should assess whether they are worth their while in terms of the subsequent research to be conducted automatically. The reason why Barbera focusses on those procedures is his keen interest in linguistic research, as proven by the remark "as is often the case", although it should be remembered that a corpus can also be used for purposes that—albeit based on language—can hardly be considered linguistic research, such as content analysis and information retrieval. However, regardless of the final goal of the analysis, the use of language (spoken or written) as raw material entails the need to take into account linguistic factors and criteria to prepare a corpus for processing. This need is apparent with reference to tokenisation and mark-up, the two early stages in corpus preparation mentioned by Barbera, but (socio)linguistic factors come into play also in the preliminary choices leading to the selection of the texts to be included in a corpus (see Sect. 7.2).

As already mentioned, tokenisation is aimed to identify the minimum units of the corpus, i.e. what we usually call "the words". It may seem a relatively straightforward procedure: the software needs to be instructed on which characters should be considered letters forming tokens and which should be considered spacers, including apostrophes, hyphens and the like. However, a researcher should be aware of the differences in terms of their use and potential consequences: should *online, on-line* and *on line* be treated alternatively as one word and two words, or should they be reduced to a standard form? Even blanks create problems: in the sentence "I give up", should *give* and *up* be counted as one verb or two distinct tokens?

Corpus mark-up includes suprasegmental information, i.e. information somehow exceeding the linear sequence of tokens in texts. This includes data which may be considered "external" to the text (often called *metadata*), such as author, title, genre, chapters, paragraphs, pages, as well as "internal" data, such as philological information (additions and cancellations, writing style such as gothic or italics, but also prose, verse, etc.) and part-of-speech (POS) tagging and lemmatisation. Since the former are of interest mainly for the purpose of strictly linguistic analysis, only POS-tagging and lemmatisation will be dealt with here under Sect. 7.3.3. However, that information is vital with a view to corpus design.[1]

7.2 What's in a Corpus? Corpus Design and Sociolinguistics

In the simplest cases, a corpus comprises all items belonging to a given class (i.e. we have an *exhaustive* corpus). For example, I can include all answers given to question n. 5 of my questionnaire, but even in this case I would probably need to compare and classify results according to the informants' gender, age, social class, etc. So I would need to consider factors of variation that are external to the corpus.

[1] Consistently with the studies illustrated in this book, all the examples provided in this chapter will be mostly in English and Italian.

Again, a corpus might include all of Shakespeare' plays, but in addition to internal factors (such as comedy vs. tragedy or problem plays vs. historical vs. fantasy plays, Medieval vs. Roman period), the aim of the research might also call for external touchstones, such as the vocabulary, style, imagery and themes used by his contemporaries in England or other countries to decide which features of Shakespeare's style are only his own and which are a cultural product of his time.

However, including all the texts belonging to a given class is not always possible: in this case, first of all researchers need to decide what type of texts should be included (i.e. the various components of the corpus) and their size; then they should identify the sources from which the desired texts can be retrieved and, finally, build the corpus. In practice, this procedure is often quite complex.

To begin with, practical issues may impact the retrieval of texts, such as copyright restrictions or their availability in electronic format. The same applies to the relative size of components, which may impinge on the overall balance and representativeness of the corpus. Generally speaking, although the same corpus may be used for a wide range of different purposes, its size and design will both mirror and influence the research that can be conducted. For example, a corpus of contemporary Italian novels will tell us very little about the historical evolution of literature in Europe; a small corpus can be balanced more easily and annotated in greater detail, while a larger corpus will probably be more useful to probe macro-patterns and overall tendencies, etc. (Hunston 2008).

Our formal education leads us to believe that language is governed by a set of rules and if speakers/writers do not comply with those rules, then their utterances/texts are incorrect. Actually, things are more complicated than that. An example is provided by the use of the "singular *they*" instead of *he* to refer back to indefinite pronouns such as *nobody* and *everybody* and avoid gender bias: utterances like "everybody should take *their* (instead of *his*) books to class" have become the norm in modern English although this usage is sometimes considered incorrect by more "traditionalist" speakers, while some modern authors even resort to feminine pronouns (Swan 2016, p. 328).

We are all aware that languages change and are used in different ways by different speakers in different places: after all, as George Bernard Shaw reportedly said, "England and America are two countries separated by a common language". And we also know that we are not expected to write in the same way we speak. Sociolinguistics deals with this kind of language variation: it is the branch of linguistics studying how language changes according to time, place, social factors, etc. Consequently, researchers in corpus linguistics may benefit from the principles of sociolinguistics in assessing how suitable a given selection of texts may be to answer their specific research questions. In particular, they may find guidance in deciding whether their corpus is appropriately balanced or certain varieties of language are overrepresented.

One of the tenets of sociolinguistics is variation and any natural language is organised according to an architecture (Coseriu 1988, pp. 294–296) based on five dimensions (Berruto 1987, pp. 19–279):

- Diachronic variation, depending on the time period when a text was produced,
- Diatopic variation, depending on the geographical area,
- Diaphasic variation, mirroring differences in the situation,
- Diastratic variation, depending on the social groups to which the speaker/writer belongs,
- Diamesic variation, connected to differences in the channel of communication, the basic opposition being written vs. oral.

In the final analysis, a hard-to-define concept such as "style" may be seen as the result of the combined and simultaneous action of all dimensions of variation plus, especially in the case of literature, the constraints imposed upon the author's creativeness by the text type and genre.

7.2.1 Diachronic Variation

Saying that languages change in time may be regarded as a truism, e.g. the pronominal system of modern English has relinquished some of the components of older paradigms, as in the second persons *thou, thee, thine, you, ye, your, yours*. However, languages differ in terms of how much and how fast they change and the consequences should be taken into account when compiling a corpus. For example, today's native speakers of English, French and German virtually need to learn a foreign language if they want to read Medieval texts in their original versions, whereas Italians seem to be more lucky since their language has remained more stable in time. However, the drawback is that apparent similarities hide differences in meaning and usage liable to be misleading. The famous incipit of Dante's sonnet:

Tanto gentile e tanto onesta pare
la donna mia quand'ella altrui saluta

may be incorrectly interpreted by modern readers since *gentile* and *onesta* today mean "kind" and "honest" with no specific reference to moral virtues (moral nobility). If a corpus including modern and older texts is used to investigate word frequency, researchers ought to be aware that the same token may have been used with very different meanings in time.

Other problems may emerge in terms of alternative forms of the same word, as in *speme* and *speranza* (hope) or *virtude* and *virtù* (virtue), depending on the proximity of the variants to their Latin origin. It can be considered a special case of synonymy, since there exist two formally different (but related) words sharing exactly the same meaning, although they differ in terms of register (the variant closer to Latin being more formal). Should those words be treated as distinct tokens or not?

On the other hand, the same word can add new meanings and lose old meanings, or the new meaning may become its main sense, as happened with *progress*.

Its original (but rare today) meaning was "the action of moving forward" (late fourteenth century, as in John Bunyan's *The Pilgrim Progress*, 1678) and the idea of development and growth was added later (seventeenth century) although today it is perceived as its basic meaning (*Oxford English Dictionary*, 1933). The repercussions should be borne in mind if, as the case is in this book, the goal of the research is tracing the history of words in terms of changes in meaning and frequency of use.

Changes in time may also involve other linguistic levels such as morphology and syntax. For example, Italian (like English) has experienced changes in the use of pronouns, with traditional subject forms (*egli, ella, esso*) being partially replaced by innovative alternatives (*lui, lei, questo*), but this process is far from completed and the system is still unstable: certain text types and registers may prefer the former (e.g. legal texts) or the latter (e.g. casual conversation). Similarly, researchers point out that in the last 50 years or so Italian has undergone many changes that also involve sentence length and syntactic complexity, with an increasing tendency to resort to shorter and simpler structures also in formal and literary uses (De Mauro 2014, pp. 154–155).

All the problems illustrated above are exacerbated by the lack of readily available material in electronic format: corpora of older texts have to be created and annotated ad hoc, and this is a very time-consuming and difficult task, calling for delicate decisions (Barbera 2009).

7.2.2 Diatopic Variation

Italians are particularly aware of the geographical variation of language since local dialects are still commonly used in everyday conversation and—since no social class or geographical area produces native speakers of standard Italian—all Italians use regional varieties featuring more or less specific traits. Especially when it comes to pronunciation, unless specifically trained to become actors or TV news presenter, Italians only need to start talking to reveal whether they have grown up in the northern or southern regions (Antonelli 2010, pp. 15–52).

English is a case in point not only as a national language but also as a lingua franca. A big temptation for students of English as a foreign language and translators alike is to consider the whole world wide web as a huge corpus of texts, and "google" sentences or expressions to check whether they are correct. More often than not, anything they key in turns out to have been used somewhere at some point. The problem, of course, is that search results can be considered reliable only by checking the origin of the texts containing the search string, i.e. whether they were actually produced in an English speaking country or not.

In terms of corpus design, another consequence of diatopic variation is that mixing together texts written by Britons and Americans might not be a good idea. If this corpus composition is considered useful, then the need may emerge to instruct the software on how to deal not only with alternative spelling (realise/realize, colour/color) and different lexical items referring to the same thing (lift/elevator, motorway/

highway, football/soccer, give someone a lift/a ride) or the same words having different meanings (pants, fag), but also with morpho-syntactic idiosyncrasies such as the use of modals and auxiliaries (I don't have/haven't got) and subject-verb agreement (Germany are/is the world champion). Many examples can be easily drawn from corpora having comparable balance but including texts produced in different regions of the Anglosphere, such as Brown, Frown, LOB, FLOB, ACE and WCNZE (Romaine 2008, par. 3.1.).

These and other kinds of differences are due to diatopic variation and should be considered when compiling a corpus and differentiating its text-components. Consequences may be far reaching at all levels: from POS-tagging and lemmatisation to lexical measures (the different usage of modals and auxiliaries may lead to different results in terms of lexical richness and lexical density), as well as content analysis, which may be affected by differences in the use of the same word with different meanings or different words for the same meaning. Such variants are particularly numerous in Italian and go under the names of *geo-homonyms* and *geo-synonyms*.

7.2.3 Diastratic Variation

Diastratic variation embraces differences in language usage determined by the population group to which speakers/writers belong within society. Just like diatopic differences, the factors affecting this language variation depend on the history and characteristics of a given culture and—most importantly—do not reflect biological differences, unless specific traits are considered, such as different voice pitch levels for men, women and children. The same applies to language differences according to the speaker's social class, which are the result of education and culture rather than biology.

Gender

Differences in the ways that men and women use language have long been of interest in the study of discourse. Despite extensive theorising, actual empirical investigations have yet to converge on a coherent picture of gender differences in language. Of course, the risk is running into stereotypes, such as "women talk more and their linguistic skills are better than those of men" (but men are better at maths, etc.) as illustrated by Abby Kaplan (2016).

There has been a fair amount of research on how men and women talk and hypotheses have been put forward to explain the differences (Lakoff 1975). In general, women tend to use less swearwords and taboo expressions (Stenström 1991), more standard forms, hedges and adjectives, more words related to psychological and social processes, while men refer frequently to object properties and impersonal topics and use more complex syntax. Furthermore, women seem to focus more on interpersonal relationships, i.e. they convey meaning but at the same time manage their relationship to the interlocutors, whereas men are strictly focussed on conveying the message (Attili and Benigni 1979).

When compiling a corpus, researchers should not only consider the role played by gender in terms of the language, but also take into account the situation in which linguistic data have been collected. In fact, one may doubt that men talk less than women if reference is made to all-male conversations about football in pubs or sports talk shows on TV. In addition, pragmatic aspects also come into play and one may wonder whether the mitigation strategies emerging from corpus analysis are the effect of gender (thus proving that women are less aggressive than men) or the speakers are actually less certain about their statements (Park et al. 2016).

Age

Undeniably, if someone has been lucky enough to become elderly, his or her linguistic choices will reflect diachronic change. There are "rules" we learn at school, many instances of "say so" or "don't say so", that we comply with even if they have become outdated. When annoyed with me, my grandmother used to say *taci*; when I want my little niece to shut up, I opt for *stai zitta* (for other instances of microlinguistic biography, see Renzi 2012).

However, a novelty in the second half of the 1900s was the rise of a social group that had been previously almost neglected: the young (for an overview of the different stages of the language of young people in Italy, see Cortelazzo 1994, par. 3). On the one hand, the language of the young could be compared to that of the partially educated: since by definition the young have yet to complete their education, they cannot be expected to master the higher and more complex registers of language. On the other hand, especially after the 1950s (the film *Rebel without a Cause* was released in 1955), they started to develop their own way of talking in order to mark the distinction between them and older generations. Language is used to establish a speech community: in addition to topics (young people constantly talk about school/university, music, fashion and sex) and rhetorical preferences (they have an inclination for puns and hyperboles), studies have shown that the choice of lexis probably is the main feature of this diastratic variety (Coveri 2014).

One of the major difficulties in identifying the language of the young is its transient nature: unfortunately, we all grow old and so do our linguistic choices, and the new generation will steer clear of them and adopt new ones. However, age is another variable to be considered when compiling a corpus.

Education Levels and Social Classes

Diastratic variation is best illustrated by membership in a social group. For example, up to the second half of the 1900s, lexical and phonological preferences in English could be said to distinguish the upper, middle and lower classes (Ross 1980).

However, differences like those detected by studies on corpora and concerning the use of everyday vocabulary in Great Britain are difficult to spot in other countries, such as Italy, owing to regional variation, which implies code switching and mixing with the local dialect.

Nevertheless, the underlying principle of diastratic variation is that those who are less wealthy probably have also attained lower education levels; consequently, they are less likely to master the full range of registers available in a given language since the higher, more formal layers, especially in writing, escape their linguistic abilities. This was particularly apparent until recently in Italy owing to low literacy levels and the competing influence of local dialects, leading to a variety called *italiano popolare* (Berruto 2012). Given the growing standardisation of (at least) popular culture, which is likely to contribute to the watering down of individual expressions reflecting social classes, education levels are a more reliable explanation for other, more general differences in corpora: lexical richness, syntactic complexity, terminological appropriateness.

7.2.4 Diamesic Variation

Diamesic variation accounts for the main, inescapable classification when analysing language. As a matter of fact, there seems to be no "third way": a text is either written or spoken, i.e. the message has been produced either with ink or paper or by uttering sounds through the human speech organs. Regardless of transcription problems (which will not be dealt with in this chapter), the distinction is not so straightforward.

An example is provided by emails and "traditional" hardcopy letters. Letters tend to be more formal: when people write letters, even to close friends or relatives, they take their time, probably prepare a rough draft, then revise it, send it and start waiting patiently for a reply. That is not the case with emails: even relatively formal emails are written almost impromptu and the sender becomes impatient if he or she does not receive a reply within the next 24 h. Emails are examples of written communication sharing some of the features of oral discourse: impromptu wording and immediate feedback are typical of face-to-face oral communication. This translates into less formal language, as if the sender were talking rather than writing: netiquette always implies a reduction of the formality of language (Fiorentino 2013).

Just like not all written texts are written in the same way, not all oral texts are structured according to the same criteria, e.g. university lectures involve the careful arrangement of topics and selection of stylistic choices. Italian linguists use the term *continuum* (cline) to refer to the seamless transition opposing written and spoken texts (Berruto 1987), the prototypical extremes of this cline being impromptu dialogue on one side (i.e. at least two interlocutors, both present in the same environment, having an unplanned conversation) and formal writing on the other (i.e. a piece of writing which is not dependent on the context in which it was produced and addresses an unspecified readership).

Needless to say, diamesic distinction is vital to achieve a balanced corpus. For example, when analysing a corpus of newspaper articles, interviews are likely to display different features compared to the other texts: information is more diluted, interrogative sentences are very frequent along with the linguistic resources necessary to manage personal interaction (greetings, tag questions, etc.) and references to the time and place of the interview (Ondelli and Viale 2010).

Conclusively, M.A.K. Halliday (1989) claims that written and spoken communication are inherently different: oral texts tend to be less dense from the lexical (and semantic) viewpoints (i.e. less content words and more function words, especially for deictic and social interaction purposes) and less rich from the lexical viewpoint (i.e. a smaller amount of words are repeated more frequently); moreover, through abstraction and nominalisation, written texts tend to present a static and synoptic picture of the world, whereas spoken texts, through a more frequent use of verbs, are more likely to produce a dynamic view of the world.

7.2.5 Diaphasic Variation

Diaphasic variation deals with what is happening, i.e. the situation in which the communication is taking place. Of course "the situation" can hardly be reduced to a set of binary oppositions (such as ± written, ± educated, ± young) unless reference is made to register in the sense of formality (± formal). Owing to the terminological overlaps and differences between English and Italian and among the many linguists who have dealt with the matter, the following definitions may be useful.

- *Register* refers to formality levels, i.e. the relationship among interlocutors imposed by the situation, which translates into a more or less accurate selection of language.
- *Languages for special purposes* are related mainly to the subject matter. People can discuss biology at a conference, in the lab or at the pub, consequently the subject matter (i.e. terminology) may interact with the formality level (Cortelazzo 1990).
- *Text types* refer to the basic cognitive functions performed by human beings through language (e.g. describing something, telling a story, supporting an argument), which translate into narrative, descriptive, argumentative texts, etc. (Wehrlich 1982).
- *Genres* account for text structures as a product of historical development in addition to the pragmatic goal of communication (Swales 2004). For example, research articles as a genre are structured in a certain way because they are used to report on research experiences, but they are also a historical and cultural product, as shown by the fact that the way research articles are written in many countries in continental Europe is heavily dependent upon the Anglo-American model involving constraints on what information should be present and how it

should be presented in the text (a parodistic application is the automatic generator of computer science research papers available on http://pdos.csail.mit.edu/scigen/).

When compiling a corpus, researchers ought to be aware of the linguistic features they can expect to find during the analysis. For example, if the corpus includes court judgments, legal terminology may be expected along with traits typical of high registers (e.g. impersonal structures, abstract and formal vocabulary, complex syntax); moreover, since judgments as a genre are composed of different text types (descriptive, narrative, argumentative, prescriptive), if samples are extracted from the full texts for the purpose of the analysis, deciding where they are taken is paramount because they may produce different data (e.g. more connectors or verbs in the past tense: see Ondelli 2013).

Diaphasic variation is called upon to account for all the factors illustrated above. Although it is probably the most obvious level of linguistic change (we are all aware that we use different words when we speak about different subjects, as we all know that we need to speak differently with a university professor or a close friend), describing a "situation" in sociolinguistic terms is a rather complex task.

Finally, there are "types of language" we are quite familiar with, such as the language of newspapers, TV or advertising (or even the language of the media in general), which cannot be identified by the subject matter (TV, the press and advertisements deal with almost anything) or genres (there are many different types of newspaper articles: feature articles, editorials, interviews, etc.) or registers (popular and quality newspapers are different). Rather, other factors come into play, such as time constraints and readership fidelity (newspapers), the imitation of spoken language and the presence of videos (TV), the pragmatic objective of "selling", i.e. convincing the audience that a given product is of unmatched quality (advertising). Language changes accordingly and researchers should consider the impact on their corpora if they decide to include texts belonging to those varieties.

7.3 Preparing a Corpus: Pre-processing

After deciding what kind of texts should be included in the corpus and retrieving the necessary linguistic data in electronic format, the next step involves preparing the corpus for processing. This means that additional choices must be made to instruct the software on how to subdivide texts into smaller units: once again, this stage is heavily dependent on the objectives of the research. The term *pre-processing* will be used here to refer to all operations carried out (semi)automatically before the actual linguistic analysis of the corpus begins. Since the following paragraphs do not deal with the matter in detail, reference will be made to the basic operations described in the operating manuals of software such as *Taltac*[2] (www.taltac.it) and *Treetagger* (http://www.cis.uni-muenchen.de/~schmid/tools/TreeTagger/).

7.3.1 Tokenisation

As already mentioned in Sect. 7.1, tokenisation basically consists in telling the software on what grounds it should decide where words (or tokens) begin and end. This may seem a straightforward task: word-token boundaries are marked by blanks or punctuation marks. Although effective, this approach can be considered naïve for several reasons.

To begin with, in ideographic languages, such as Chinese, this rule of thumb does not apply since texts do not provide information about word boundaries and tokenisation proves very difficult. However, also alphabetic languages, like English and Italian, pose a wide range of problems. An example is provided by the way many commonly used word-processors count words in documents, so that a strings of characters such as *l'amore è una cosa meravigliosa* and *love is a wonderful thing* will both produce a word count of five, although an Italian reader would strongly object to the idea of *l'amore* being one single word, in the same way an English speaker would probably disagree with a word count of four for the sentence "I can't do it". An obvious solution could be to consider apostrophes as word boundary markers, or even instruct the software to replace the contracted form (so that *he doesn't* becomes *he does not*, *wie geht's?* becomes *wie geht es*?) but then one should decide what to do with cases such as the colloquial negative verb *ain't* in English, since no words such as *ain* or *ai* exist.

This kind of decisions will obviously impact all measures based on the number of tokens in a corpus, such as lexical richness. In Italian, is the sentence *puoi dirlo forte* (you can say it loud) composed of three or four tokens, considering that it can be rephrased as *lo puoi dire forte*, with the clitic pronoun separated from the other words? And how could the software distinguish *dirlo* from *Carlo* (Charles)? Should the French interrogative verb-subject structure (*voulez-vous danser*?) be de-hyphenated? What about multiple options such as *on line, on line* and *online*?

The seemingly straightforward "no blank and no punctuation within words" rule proves ineffective when it comes to numbers (100,000 or 100 000 are variants) and—more significantly—multiword expressions. Proper names like *New York, The Hague, L'Aquila, Città di Castello, Wall Street, The Financial Times,* etc. can be recognised through more or less exhaustive lists (named-entity recognition), and the same possibly applies to complex grammar words such as prepositions, conjunctions and even adverbs (*because of, as well as, for sure, in the light of*). But what about content multiword strings, such as *credit card, phone booth, floppy disk, tug-of-war, smart TV,* which make up a relevant share of the neologisms that emerge constantly in languages and, consequently, cannot be included in closed lists? As shown by the examples provided so far, tokenisation can be difficult, and even more so in languages particularly prone to the formation of more or less stable compound words such as German: *Haustür* becomes *house door* in English and *porta di* (or *della*) *casa* in Italian; *Windgeschwindigkeit* is translated as *wind speed/velocità del vento*; *Freudentränen* (with a morphological modification of *Freude* into *Freuden*) means *tears of joy/lacrime di gioia* and one may wonder whether they should be treated as distinct lexemes or stripped down to their components.

Another typical feature of languages like English and German is the use of verbs including particles that modify their basic meaning (like *get up, get down, get off, get along, aufmachen, zumachen,* etc.) and sometimes can move along the sentence away from the verb itself: "I've come to *cheer* you *up*"; the basic verb *mitteilen* is separated when conjugated, as in "wir *teilen* Ihnen das Ergebnis *mit*" (we notify the result to you). To a lesser extent, Italian can produce similar constructs in the case of pronominal verbs (*alzarsi* = to get up, but *mi* sono *alzato* = I got up) and phrasal verbs (*andare su, giù, sopra, sotto, a male,* etc.). By this stage, it is obvious that deciding what is a word (or token) before analysing a corpus is vital: set phrases like *to kick the bucket* or *prendersela comoda* (i.e. "to take it easy", which splits into *me la prendo comoda* when conjugated) can lead to a sharp increase or decrease in the word count based on how they are considered.

Depending on the languages included in a corpus and the software used, different approaches can be implemented but no overall solution is available: decisions must be made and, according to them, results will eventually have to be interpreted in one way or another. For example, the Italian resources available in *Taltac*[2] can recognise a large number of proper names, municipalities, newspapers and famous personalities and also many multiword expressions. However, *Treetagger*, probably one of the most commonly used POS-tagging programmes freely available for a number of languages, cannot recognise modal and auxiliary verbs, so that all compound forms are treated as distinct lexical items.

At a more advanced stage of the analysis, if those aspects are neglected and— for example —the analysis aims to measure the presence of static or dynamic expressions in a corpus, problems may emerge in interpreting the results if all the verbs of movement used to convey the future tense (*I am going to eat/Je vais manger*) are not recognised during pre-processing and consequently expunged. Finally, when a corpus is to be compared with other reference corpora (e.g. is a corpus of texts written by individuals with autism more/less lexically rich than texts written by university students or newspaper articles or whatever?), if researchers are not sure that the same tokenisation procedures have been implemented, results may be completely misleading.

7.3.2 *Punctuation*

Punctuation marks need to be considered to identify sentences, clauses and tokens. A problem which will not be dealt here is the amount of incorrect punctuation and orthography frequently found in texts retrieved from the Internet or scanned by means of OCR programmes. Sometimes, punctuation marks or blanks are missing and run the risk of impinging upon the results of sentence length calculations (e.g. in the case of missing full stops at the end of sentences) or tokenisation (e.g. in strings like "time.He" or "coming?However").

Punctuation can actually occur within words, e.g. hyphens as in *pre- and post-war period*, parentheses as in *evening(s)*, slashes as in *Studenten/Innen*, as well as full stops (especially in abbreviations such as *U.S.A, fig.*) and exclamation marks (*Yahoo!*). Additional cases should be considered according to the language of the corpus (e.g. ordinal numbers are followed by full stops in German).

Problems deriving from how punctuation is treated include capitalised words and sentence boundaries. *Some* at the beginning (as in "Some think you are right") or in the middle of a sentence (as in "I would like to have some more") will be considered as different forms unless the software is instructed to de-capitalise words occurring at the beginning of sentences which do not belong to the list of proper names (this aspect is even more complex in German, in which all nouns are capitalised). However, deciding where sentences begin and end is difficult, since publishers do not always comply with strict rules when, e.g., a text contains direct speech. All the following alternatives can be found in Italian (Mortara Garavelli 1985):

- *Marco mi ha detto: "Domani arriverò tardi, non mi aspettare". Allora sono andato a dormire.*
- *Ieri ho incontrato Marco – Ci vediamo domani – ha esclamato.*
- *Rientrai a casa tardi ma i miei mi stavano aspettando – Dove sei stato fino ad adesso? – Sono affari miei - e corsi in camera.*

Establishing sentence boundaries is necessary to calculate average sentence length, which is used as an index of presumed syntactic complexity and text readability (in the case of Italian, see the Gulpease index illustrated on www.corrige.it).

7.3.3 POS Tagging and Lemmatisation

As shown above, after compiling a corpus, its preparation for the analysis can hardly be defined "automatic" even for a relatively simple task such as tokenisation. The next steps in the preparation of a corpus are mutually related and have a great impact on any subsequent analysis: assigning tokens to POS categories, i.e. deciding whether they are adverbs, verbs or nouns, and—consequently—lemmatising, i.e. identifying the basic forms used as entries of a dictionary, which means (for many languages) the singular masculine nominative for pronouns and nouns, the infinitive for verbs, etc.

In assigning the grammatical function of tokens, morphology and syntactic position play important—and generally inversely proportional—roles. Morphologically poor languages such as English tend to comply with a fixed word order, whereas morphologically richer languages like Italian allow greater flexibility. However, in the first case, the price to pay is greater ambiguity: virtually any word in English can function as a noun, an adjective (modifier) or a verb, as in:

- *Love is important in everybody's life.*
- *Italians love pasta.*
- *Our love affair is over.*

Inflectional languages, such as Italian, are characterised by greater morphological variation and can display the role assigned to a word overtly through suffixes; e.g. the *–vo* ending in *amavo* immediately shows that it is a verb in the first person singular of the imperfective past tense. Of course, this does not mean that ambiguity is ruled out: *andiamo* can be the first person plural of the indicative, subjunctive and imperative moods; *potere* can be either a noun (power) or an infinitive (can); *letto* is either a noun (bed) or a past participle (read), etc.

The results produced by tagging programmes are not error free, the error rate being dependent on training data. Some programmes, like *Taltac*[2], provide a list of options from which researchers can decide how each token should be disambiguated. This approach can be very time-consuming in the case of large corpora since it calls for checking the contexts of all items with undefined POS-tag. Furthermore, even if a researcher decides to go through the painstaking task of manual tagging, difficulties may still emerge since the status of certain past participles as adjectives or verbs or the classification of modal verbs are a debatable matter also for linguists: additional examples are provided by a simple word such as *not* in English or the innumerable roles played by the form *che* in Italian.

Lemmatisation is directly connected to POS-tagging: only when the grammatical role of a word-token is established can its basic form be ascertained. Especially in languages using different forms to express morphological attributes such as case, number, gender, tense and mood the purpose of lemmatisation is overcoming the limits set by the contingent (contextual) nature of lexical choices, e.g. in Italian, masculine, feminine, singular and plural variants of nouns, determiners and adjectives; three persons (first, second, third), either singular or plural, and a wide range of moods and tenses in the case of verbs, not to speak of the cases of personal pronouns. Only after having established that a group of tokens are past participles can the software assign the variants *letti, lette, letta, letto* to the lemma *letto.* This operation is quite complex owing to the ambiguity deriving from homographs (i.e. words with identical spelling but different meanings and/or syntactic roles): in modern Italian not only *principi* may be the plural form of *principe* (prince) or *principio* (principle), but a form like *stato* can be a noun (*state*) or the past participle of either the verb *essere* (to be) or *stare* (to stay), with all their possible variants in terms of gender and number (*stato, stata, stati, state*).

The problems posed by POS-tagging and lemmatisation can be exacerbated by semantic ambiguity and figurative speech. In certain contexts, even human beings may be uncertain about the syntactic role and meaning to be assigned to a token (as in "we saw her duck"). In the case of (especially stereotypical) figurative speech (*he has a sweet tooth; hai la testa tra le nuvole*), one may wonder whether it makes sense to consider each token (and relevant lemma) individually or treat these data as a multiword expression to be lemmatised as such.

However, POS-tagging and lemmatisation are important not only to calculate straightforward measures of lexical complexity of a corpus, such as lexical density, but also to achieve a finer classification of texts. For example, the frequency of verbs in the past tenses may reveal the presence of narrative text types, while the incidence of the gerund, subjunctive and conditional moods—in combination with average sentence length—may point to greater syntactic complexity. Finally, although in corpus linguistics lexical richness is traditionally based on forms, one may object that that is not what is commonly meant when a person is said to have "a rich vocabulary". What is generally implied in this case is not that a person can derive many inflectional variants from the basic lexical forms, but that he or she controls and uses a wide range of synonyms; in other words, reference is made to the number of lemmas, and not of inflected forms.

7.3.4 Stemming

Stemming may be considered an automatic approach to lemmatisation (Fitschen and Gupta 2008, par. 2) which is not based on a pre-existing list of forms (lexicon). The lemmatisation process illustrated in Sect. 7.3.3 above aims to reduce all inflectional variants to a basic form which is relatively arbitrary: Italian dictionaries list verbs according to their infinitive forms, whereas Latin dictionaries opt for the first person singular of the present indicative.

Stemming also reduces inflectional variants but this result is achieved by truncating word forms and discarding the remaining material without attempting any morphological identification. Truncation is performed according to an algorithm that reduces the string of characters regardless of the linguistic material being eliminated: it may consist of suffixes (such as *–ed* in *rounded*) or full lexical components in compound words (such as *-about* in *roundabout*). The final result may be linguistically motivated (as in the previous examples) or not motivated, as is the case with the truncation of *timing* leading to *tim.*

Albeit useful to reduce words to a common representation for information retrieval purposes, stemming poses a number of different problems in terms of the selection of stemming rules and their application to languages envisaging the insertion of morphological material in the word form string (e.g. infixes in Turkish).

7.4 Conclusion

In the final analysis, the main difficulties in tokenisation and lemmatisation emerge at a level requiring an analysis that exceeds mere formal traits and calls for assessing the contexts in which tokens are included to appraise their syntactic function and meaning. Additional operations have been devised with a view to corpus analysis, envisaging the (partially automatic or hand-made) syntactic and semantic tagging of texts beyond the POS level, e.g. treebanks (Sampson 2003) and pragmatic annotation (McEnery and Wilson 2001). Expanding on such approaches would exceed the scope of this chapter, which is only meant to provide a short overview of the main issues emerging in respect of corpus design and pre-processing.

A corpus is made of texts, which in turn are composed of word-tokens, i.e. the data used for automatic and statistical analysis. However, since no corpus, no matter how large it is, will ever be capable of accounting for all the possibilities within a natural language, the logical consequence is that, in order to decide whether a corpus is "good" or not, whether it can be useful and reliable, the invariable touchstone is the purpose for which it has been compiled. Of course, the argument can be reversed: a good corpus can only be built with clear research questions in mind. Consequently, a full account of all the choices made in terms of corpus design and preprocessing should be illustrated explicitly so that comparisons are possible with previous or subsequent studies and the corpus may be fruitfully exploited for other research projects.

Even when dealing with exhaustive corpora, like those is in this book, questions must be posed about their representativeness and reliability. Their classification according to sociolinguistic factors can reveal much about the validity of the results of the analysis: what is the time period considered? What topics, text types (written or oral) and genres (addressing specialists or the general public) have been included? What is the geographical origin of the texts? What is the inner balance of the subcorpora (if any) included in the general corpus? What tokens of language were considered and how were they identified? Only by answering those questions can conclusions be drawn and predictions be made about the level of generalisation of the findings.

References

Antonelli, G. (2010). Lingua. In A. Afribo & E. Zinato (Eds.), *Modernità italiana. Cultura, lingua e letteratura dagli anni Settanta a oggi* (pp. 15–52). Roma: Carocci.

Attili, G., & Benigni, L. (1979). Interazione sociale, ruolo sessuale e comportamento verbale: lo stile retorico naturale del linguaggio femminile nell'interazione faccia a faccia. In F. A. Leoni & M. R. Pigliasco (Eds.), *Retorica e scienze del linguaggio: atti del 10. Congresso internazionale di studi, Pisa, 31 maggio - 2 giugno 1976. SLI, Società di linguistica italiana* (pp. 261–280). Roma, Bulzoni.

Barbera, M. (2009). *Schema e storia del Corpus Taurinense: linguistica dei corpora dell'italiano antico*. Alessandria: Edizioni dell'Orso.

Barbera, M., Corino, E., & Onesti, C. (2007). Cosa è un corpus? Per una definizione più rigorosa di corpus, token, markup. In M. Barbera, E. Corino, & C. Onesti (Eds.), *Corpora e linguistica in rete* (pp. 25–88). Perugia: Guerra.

Berruto, G. (1987). *Sociolinguistica dell'italiano contemporaneo*. Roma: La Nuova Italia Scientifica.

Berruto, G. (2012). L'italiano popolare e la semplificazione linguistica. In G. Berruto (Ed.), *Saggi di sociolinguistica e linguistica* (pp. 141–181). Alessandria: Edizioni dell'Orso.

Cortelazzo, M. A. (1990). *Lingue speciali. La dimensione verticale*. Padova: Unipress.

Cortelazzo, M. A. (1994). Il parlato giovanile. In L. Serianni & P. Trifone (Eds.), *Storia della lingua italiana, vol. II, Scritto e parlato* (pp. 291–317). Torino: Einaudi.

Coseriu, E. (1988). *Einführung in die Allgemeine Sprachwissenschaft*. Tübingen: Francke.

Coveri, L. (2014). *Una lingua per crescere. Scritti sull'italiano dei giovani*. Firenze: Franco Cesati editore.

De Mauro, T. (2014). *Storia Linguistica dell'Italia repubblicana dal 1946 ai nostri giorni*. Roma-Bari: Laterza.

Fiorentino, G. (2013). *Frontiere della scrittura: lineamenti di web writing*. Roma: Carocci.

Fitschen, A., & Gupta, P. (2008). Lemmatising and morphological tagging. In A. Lüdeling & M. Kytö (Eds.), *Corpus Linguistics. An International Handbook* (pp. 552–564). Berlin: Walter de Gruyter.

Halliday, M. A. K. (1989). *Spoken and written language*. Oxford: OUP.

Hunston, S. (2008). Corpus compilation and corpus types. In A. Lüdeling & M. Kytö (Eds.), *Corpus Linguistics. An International Handbook* (pp. 154–168). Berlin: Walter de Gruyter.

Kaplan, A. (2016). *Women talk more than men ... and other myths about language explained*. Cambridge: Cambridge University Press.

Lakoff, R. (1975). *Language and Woman's Place*. New York: Harper.

McEnery, T., & Wilson, A. (2001). *Corpus Linguistics: An Introduction* (2nd ed.). Edinburgh: Edinburgh University Press.

Mortara Garavelli, B. (1985). *La parola d'altri: prospettive di analisi del discorso*. Palermo: Sellerio.

Ondelli, S. (2013). Un genere testuale attraverso i confini nazionali: la sentenza. In S. Ondelli (Ed.), *Realizzazioni testuali ibride in contesto europeo. Lingue dell'UE e lingue nazionali a confronto* (pp. 67–92). Trieste: EUT.

Ondelli, S., & Viale, M. (2010). L'assetto dell'italiano delle traduzioni in un corpus giornalistico. Aspetti qualitativi e quantitativi. *Rivista internazionale di tecnica della traduzione, 12*, 1–62.

Oxford English Dictionary (1933). Oxford: OUP.

Park, G., et al. (2016). Women are Warmer but No Less Assertive than Men: Gender and Language on Facebook. *PLOS, 25*(2016), e0155885. https://doi.org/10.1371/journal.pone.0155885.

Renzi, L. (2012). *Come cambia la lingua: l'italiano in movimento*. Bologna: il Mulino.

Romaine, S. (2008). Corpus linguistics and sociolinguistics. In A. Lüdeling & M. Kytö (Eds.), *Corpus Linguistics. An International Handbook* (pp. 97–111). Berlin: Walter de Gruyter.

Ross, A. S. C. (1980). U and non-U. In N. Mitford (Ed.), *Noblesse oblige* (pp. 11–38). London: Futura.

Sampson, G. (2003). Thoughts on Two Decades of Drawing Trees. In A. Abeillé (Ed.), *Treebanks* (pp. 23–41). Dordrecht: Springer.

Stenström, A.-B. (1991). Expletives in the London-Lund Corpus. In K. Aijmer & B. Altenberg (Eds.), *English Corpus Linguistics: In honour of Jan Svartvik* (pp. 230-253). London: Longman.

Swales, J. M. (2004). *Research Genres: Explorations and Applications*. Cambridge: Cambridge University Press.

Swan, M. (2016). *Practical English Usage*. Oxford: OUP.

Wehrlich, E. (1982). *A Text Grammar of English*. Heidelberg: Quelle & Meyer.

Chapter 8
Automatic Multiword Identification in a Specialist Corpus

Pasquale Pavone

Contents

8.1 Introduction 152
8.2 Related Work 153
8.3 Lexical-Textual Model for Compiling a Terminological Dictionary 153
  8.3.1 Grammatical Tagging 154
  8.3.2 Syntactic Structures of Nominal Syntagmas 154
8.4 The Corpus of Philosophy Journals 156
  8.4.1 The Journals 156
  8.4.2 Out-of-Context English Grammatical Tagging 158
  8.4.3 Searching the Syntactic Structures in the Corpus 160
8.5 Analysis of Semantic Dimensions 161
8.6 Conclusions 165
References 165

Abstract In a logic of study of specialist-technical corpora, this work proposes the definition of a lexical-textual model for the automatic identification of the nominal Multiword Expressions present in texts. In automatic text analysis, particular attention usually devoted to recognizing the nominal Multiword Expressions in a corpus, which include both nominal idiomatic expressions and linguistic collocations. This vast class of Multiword Expressions includes technical terms and compound personal nouns. They are thus often found in specialist-technical language. Though they are not nominal idioms, these complex lexemes represent technical or specialist expressions. Accurate detection of Multiword Expressions enables us to disambiguate the meaning of words and to define or enhance terminological glossaries for a specific specialist sector. Our objective is reached through the recognition of the syntactic structures which define the nominal expressions. Multiword Expressions represent the universe of disambiguous subjects and objects in a text, that is to say, the terminology of the discourse. It is shown how the use of factor analysis in a

P. Pavone (✉)
Università degli Studi di Modena e Reggio Emilia, Modena, Italy
e-mail: pasquale.pavone@unimore.it

A. Tuzzi (ed.), *Tracing the Life Cycle of Ideas in the Humanities and Social Sciences*, Quantitative Methods in the Humanities and Social Sciences,
https://doi.org/10.1007/978-3-319-97064-6_8

limited number of Multiword Expressions is able to rebuild the same structure of the whole vocabulary in analysis. The procedure here presented is applied to the corpus of documents made of a collection of titles of papers published in the journals *Mind*, *The Monist*, *The Journal of Philosophy* and *The Philosophical Review*, from their foundation to the last number for 2016.

Keywords Multiword expressions · Technical language · Part-of-speech tagging · Regular expressions · Taltac2 software

8.1 Introduction

This chapter presents a lexical-textual model (Pavone 2010) which, by defining syntactical structures, can automatically identify and extract Multiword Expressions (MWEs) in texts with the support of Taltac software.[1] In automatic text analysis (Bolasco 2013), particular attention usually devoted to recognizing the nominal MWEs in a corpus (Sag et al. 2002), which include both nominal idiomatic expressions and linguistic collocations. Accurate detection of MWEs enables us to disambiguate the meaning of words and to define or enhance terminological glossaries for a specific specialist sector.

An idiomatic MWE is a complex lexeme characterized by being non-compositional the meaning of the expression does not transparently follow from the meaning of the words comprising it (Elia 1995). Collocations, on the other hand, are sequences of two or more words with a strong mutual association (Sinclair 1991). Both idiomatic MWEs and collocations are often found together on syntagmatic axes. This vast class of MWEs includes technical terms (Justeson and Katz 1995) and compound personal nouns. They are thus often found in specialist-technical language. Though they are not nominal idioms, these complex lexemes represent technical or specialist expressions (De Mauro 1999–2007).

An MWE is less polysemous than mono-words and accurate MWE detection should lead to a nontrivial improvement in Word Sense Disambiguation (Agirre and Edmonds 2007). Identifying and lexicalizing the more common MWE structures makes it possible to disambiguate the objects and subjects of texts, which are the central element of the message conveyed in a discourse. Specifically, identifying them enables us to define the terminology in technical-specialist corpora and select the lexical units of analysis (word and multiword) for further automatic analysis of textual data in order to represent the information contained in it.

[1] Taltac2 stands for Automatic Lexical and Textual treatment for Analysis of the content of a Corpus, developed from research carried out at the University of Rome La Sapienza (Bolasco 2010).

8.2 Related Work

Natural Language Processing (NLP) and Computational Linguistic studies offer numerous methods for detecting MWEs. A first possibility consists of detecting MWEs using an external glossary, but obviously the main limitation of this method is that the expert dictionary cannot retrieve all of the MWEs actually present in the text.

A great variety of automatic methods are used, depending on the data, the particular tasks and the types of MWEs to be extracted. The most frequently used metrics, inter alia, are Mutual Information (MI), (Church and Hanks 1990), *t*-score (Church et al. 1991), log-likelihood (Dunning 1993), and significance index (IS) (Morrone 1993). The different indexes assign a value to the different segments of words produced by segmenting the corpus in sequences of 2, 3..., *n* words which are repeated again and again in the text (Salem 1987) between two obvious separators, i.e., between punctuation marks. For example, the Mutual Information Index compares the probability of observing two words, x and y, together (the joint probability) with the probabilities of observing x and y independently (chance). If there is a genuine association between x and y, then the joint probability $P(x, y)$ will be much larger than chance $P(x)$ $P(y)$, and consequently $I(x, y) \gg 0$. The word probabilities $P(x)$ and $P(y)$ are estimated by counting the number of observations of x and y in a corpus, $f(x)$ and $f(y)$, and normalizing by N, the size of the corpus.

The common disadvantage of these methods is their dependency on the number of words included in the MWE. Although a large number of studies use MI for bigram extraction, only a few employ the MI measure for three or more collocates. In all cases, indexes allow us to make a graduated list of the repeated segments thus identified, and MWEs are selected manually from the ranking offered by the indexes, rather than being extracted automatically.

The procedure presented here for identifying and extracting MWEs is totally automatic, as it formalizes their syntactic structures using Regular Expressions and is based on meta-grammatical information for the vocabulary under analysis.

8.3 Lexical-Textual Model for Compiling a Terminological Dictionary

The proposed algorithm entails three successive stages: grammatical tagging of the words (types[2]) of the vocabulary; definition of the morfo-syntactic structures and searching the bodies of interest in the corpus; MWE lexicalization.

[2]Types are the different entries of the Vocabulary. Each presence of a type in the Corpus is an occurrence.

8.3.1 Grammatical Tagging

Automatic grammatical annotation of the various words is quite a complex process for which there are two possible alternatives: "out-of-context" grammatical tagging and "in-context" grammatical tagging (Lenci et al. 2005). The first method allows us to annotate the words of the vocabulary under analysis using linguistic resources. To do so, we proceed with a comparison between the words of the vocabulary in a reference dictionary which contains "static" grammatical information, i.e., information independent of the context in which the words occur. In this way, only those words which present a unique non-ambiguous grammatical category are noted. The output from this method of annotation is a word which is grammatically tagged in a very precise way (no false positives), but with a large number of ambiguous annotations (with false negatives) because of the different grammatical classes which can be associated with the same word.

The second automatic grammatical annotation method consists of formulating algorithms which analyze the words in the context in which they occur. Such algorithms are generally based on linguistic rules and statistical probabilistic models (Grigolli et al. 1995; Schmid 1994, 1995), and can assign a certain grammatical category to all the vocabulary entries, though with a variable margin of error. This second method's output is a completely annotated vocabulary (no false negatives), but inevitably a certain number of words will be erroneously tagged (with false positives).

Although both methods are valid, the model we propose uses out-of-context grammatical tagging. When searching for nominal syntagma formed essentially by combinations in sequence of adjectives and nouns (see below for details), it is possible to use the multiple annotation of the ambiguous words in the algorithm without finding false positives among the MWEs.

8.3.2 Syntactic Structures of Nominal Syntagmas

In the second stage, the algorithm identifies the syntactic structures which allow us to retrieve the nominal syntagmas in the text.

To identify the grammatical structures from which nominal syntagmas can be extracted, we started from the nominal idioms in the Dictionary of Polyforms, a linguistic resource provided in Taltac2 software.

This resource was obtained by determining the occurrence of the idioms which have compound nouns (Elia 1995, 1996) in a standard Italian dictionary (Bolasco and Morrone 1998).

We found that 60% of the nominal polyforms in the reference resource consist of <N + A> structures (for Italian, translatable as <A + N> in English) or <N + N> structures, while 32% consist of an <N + PREP + N>, structure, of the sequence <N + *of* + N>. The added value of this last structure is provided by the preposition

<of> which, as it denotes property (Rouget 2000), introduces the second noun and adds a further meaning to the syntagma. This structure firmly links the two nouns together, recognizing them as the head and the modifier, or both heads of a nominal collocation (Sinclair 1991).

These structures are thus applied as textual queries for the extraction of complex words. In the experimental and evaluation stages for these REs, it was found that the structure:

$$\mathrm{N} + of + \mathrm{N} \tag{8.1}$$

can be repeated in the structure:

$$\mathrm{N} + of + \mathrm{N} + of + \mathrm{N} \tag{8.2}$$

or developed in the structure:

$$\ll \mathrm{A} + \mathrm{N} > +of + \mathrm{N} > \tag{8.3}$$

$$< \mathrm{N} + of + < \mathrm{A} + \mathrm{N} \gg \tag{8.4}$$

Furthermore, when the longer structures (8.2), (8.3), and (8.4) generate recognized and lexicalized expressions, any false positives which may be produced are automatically eliminated in the subsequent lexicalization of the shorter expressions obtained by (8.1).

If, for example, we use (8.4) and find the longer expression *<elements of physiological psychology>* or *<philosophy of common sense>* in the corpus, the nonsensical expressions *<elements of physiological>* and *<philosophy of common>* obtained by (8.1) will automatically not be recognized, as they are always included in structure (8.4). On the other hand, the collocations *<physiological psychology>* and *<common sense>* remains as units of analysis in the dictionary, as their use in terms of occurrence goes beyond the specifications of *<elements>* and *<philosophy>* (first nouns).

By contrast, applying REs to identify the structure <A + N> can generate a sizable number of false positives if the REs are not refined. As a large number of false positives is yielded by the association between a noun and a determinative adjective, we consider only qualifying adjectives.

Once the individual REs have been validated in the exploratory stage of the analysis, we define a single textual meta-query (role model) (Bolasco and Pavone 2010) which brings together the single $F(x)$, thus completing the model. However, such a vocabulary of entities may present false positives which are refined, first by excluding those words which have fewer than 5 occurrences, and then by lexicalizing the recognized entities by order of extension, from the longest to the shortest. At this point, the lexicalized entities are assumed as a "meta-dictionary" (model as available resource).

In Sect. 8.4, we illustrate the application of the lexical-textual nominal MWE recognition model to the corpus of philosophy journals, consisting of 32,654 titles of papers published in four different journals since their foundation until 2016. The entire analysis was carried out using Taltac2 software.

In Sect. 8.5, we demonstrate how a selection consisting only of MWEs can, in factorial analysis, reconstruct the same categorical structure generated within the whole corpus.

8.4 The Corpus of Philosophy Journals

The corpus of documents used to apply the model consists of a collection of titles of papers published in four philosophy journals: *Mind, Monist, Journal of Philosophy* and the *Philosophical Review*, from their foundation to the last number for 2016. The corpus consists of 32,654 titles. The first stage in the analysis consists of structuring the corpus's textual information in the Vocabulary Data Base and Documents Data Base, which define the lexical setting and the textual setting, respectively. For this purpose, we introduce the corpus into a Taltac2 work session. After initial word normalization, the corpus includes a total of 193,173 occurrences (Table 8.1).

8.4.1 The Journals

Mind[3]

Mind is a quarterly peer-reviewed academic journal published by Oxford University Press on behalf of the Mind Association. Having previously published exclusively philosophy in the analytic tradition. Mind has long been a leading journal in philosophy. For well over 100 years, it has published the best new work in all areas of the subject. The journal continues its tradition of excellence today. The journal aims to take quality to be the sole criterion of publication, with no area of philosophy, no style of philosophy, and no school of philosophy excluded. Each issue also contains a selection of book reviews that summarize and evaluate some of the most

Table 8.1 Type count and occurrences in journals

Journal	Type count	Occurrences
Mind	6985	48,866
Monist	4603	21,886
Philosophical Review	7172	54,636
Journal of Philosophy	9917	67,785

[3] https://academic.oup.com/mind

interesting recent publications in the discipline. Mind has always enjoyed a strong reputation for the high standards established by its editors and receives over 600 submissions each year. The editors seek advice from a large number of expert referees, including members of the network of associate editors and the Editorial Board.

The Monist[4]

Founded in 1888 by Edward C. Hegeler, The Monist is one of the world's oldest and most important journals in philosophy. It helped to professionalize philosophy as an academic discipline in the United States by publishing philosophers such as Lewis White Beck, John Dewey, Gottlob Frege, Hans-Georg Gadamer, Sidney Hook, C.I. Lewis, Ernst Mach, Charles Sanders Peirce, Hilary Putnam, Willard Van Orman Quine, Bertrand Russell, and Gregory Vlastos. The Monist publishes quarterly thematic issues on particular philosophical topics which are edited by leading philosophers in the corresponding fields. As a result, each issue is a collected anthology of continuing interest. The Monist is published by Oxford University Press.

The Journal of Philosophy[5]

The Journal of Philosophy is a monthly peer-reviewed academic journal on philosophy, founded in 1904 at Columbia University. Its stated purpose is "to publish philosophical articles of current interest and encourage the interchange of ideas, especially the exploration of the borderline between philosophy and other disciplines." Subscriptions and online access are managed by the Philosophy Documentation Center.

The Journal was ranked the second highest-quality philosophy journal in a poll conducted on the popular philosophy blog Leiter Reports and is widely regarded as one of the most prestigious journals in the field. The journal also publishes the Dewey, Woodbridge, and Nagel Lectures series held at Columbia University.

The Philosophical Review[6]

The Philosophical Review is a quarterly journal of philosophy edited by the faculty of the Sage School of Philosophy at Cornell University and published by Duke University Press (since September 2006). The journal publishes original work in all areas of analytic philosophy, but emphasizes material that is of general interest to

[4] http://www.themonist.com/

[5] https://www.journalofphilosophy.org/

[6] https://read.dukeupress.edu/the-philosophical-review

academic philosophers. Each issue of the journal contains approximately two to four articles along with several book reviews. In continuous publication since 1892, the Philosophical Review has a long-standing reputation for excellence and has published many papers now considered classics in the field.

8.4.2 *Out-of-Context English Grammatical Tagging*

The linguistic resources provided by Taltac2 software permit us to perform out-of-context grammatical tagging of the words in the vocabulary.

Results of lemmatization are summarized in Table 8.2. Grammatical classes are ordered by decreasing number of total occurrences. We have highlighted the classes of ambiguous words (TAG J in the Vocabulary DB CAT field) which account for 48% of the total occurrences, and of untagged types, i.e., words that were not present in any grammatical list in the linguistic resources, or 7756 lexical units accounting for 12.4% of the total occurrences. The remaining 40% of the words are clearly disambiguated.

Table 8.3 shows the first 15 words that the software did not tag. Normally, the untagged words in a corpus are proper nouns; in the case under study, which involves multilingual documents, they are mostly terms of foreign origin (German, Italian, French), and names of philosophers cited in the documents. In these cases, we tagged only the names and foreign nouns as (N).

With regard to ambiguous words, i.e., words whose grammatical category cannot be defined exclusively, Table 8.4 shows details of the grammatical ambiguities (tagged in the Vocabulary DB CAT_AC field) by decreasing number of types. As can be seen, the first four groups of ambiguous terms account for more than 95% of the words to be disambiguated, possibly by investigating concordances. Obviously, a disambiguation procedure of this kind would be excessively time-consuming, and

Table 8.2 Lexical units categorized grammatically and listed by decreasing occurrence

CAT	Type count	Occurrences	% Occ. in total
J	3071	91,966	47.7
N	4888	56,814	29.4
noTag	7756	24,126	12.4
A	918	7941	4.1
DET	4	4831	2.5
PREP	15	3797	1.9
V	279	1657	0.9
NUM	51	868	0.4
AVV	110	693	0.4
PRON	19	249	0.1
CONG	3	226	0.1
ESC	5	5	0.0

Table 8.3 First 15 untagged types

Type	Occurrences
de	813
la	777
und	534
philosophie	392
des	378
Aristotle	375
mr	304
le	227
les	209
Wittgenstein	205
Descartes	164
psychologie	153
zur	147
James	146
Dewey	123

Table 8.4 Details of the potential grammatical classes of 3071 ambiguous words previously tagged as J in the CAT field, with more than ten types

CAT_AC	Type count	Occurrences
N + V	1272	17,526
A + N	766	11,085
A + V	377	1263
A + N + V	367	2434
A + AVV + N	44	1185
A + AVV	41	373
AVV + N	25	568
A + AVV + N + V	24	370
N + PRON	13	495
A + AVV + V	10	98

we would lose the automatization advantage provided by text mining. The reader should bear in mind that this stage of grammatical tagging is instrumental to the second stage of lexical processing which defines the model used to find nominal syntagmas (see Sect. 8.3.2) where the principle grammatical classes are nouns and adjectives, and word ambiguity can be used as an element of the lexical-textual model. Thus, considering only the first five groups of ambiguous words highlighted in grey in Table 8.4 for the grammatical annotation process, we proceeded with multiple tagging for those groups in which the ambiguity affected adjectives as well as nouns, assigning the tag (N) and (A) to each of the corresponding words. Ambiguous words belonging to group (A + V) are tagged only as potential adjectives (A), while words belonging to group (N + V) are tagged only as potential nouns (N).

Table 8.5 The first 16 types in the vocabulary with forced annotation in CAT_SEM field

Types	Occurrences	CAT	CAT_AC	CAT_SEM
philosophy	2772	N	N	,N,
theory	1387	N	N	,N,
logic	928	N	N	,N,
science	797	N	N	,N,
moral	764	J	A + N	,A,N,
ethics	753	N	N	,N,
psychology	727	N	N	,N,
philosophical	659	A	A	,A,
knowledge	648	N	N	,N,
nature	637	N	N	,N,
study	593	J	N + V	,N,
history	589	N	N	,N,
mind	580	J	N + V	,N,
Kant	534	N	N	,N,
truth	525	N	N	,N,
thought	461	J	N + V	,N,

Tagged types in the vocabulary are listed in Table 8.5. The CAT column shows the grammatical tags assigned by out-of-context tagging, where (J) are the ambiguous words, while the CAT_AC column shows the potential grammatical classes for the ambiguous words. The CAT_SEM column is based on the choices explained above. Considering the ambiguities as elements of the model, i.e., using multiple grammatical categorization (in cases of <A + N>) or potential categorization (in cases <N + V>, <A + V>, or <A + AVV + N) does not jeopardize the final objective because such grammatical annotations are important for the recognition of complex forms of syntagmatic types. In this way, as suggested in Sect. 8.3.2 and as we will see below, potential false positives are excluded from the results of the exclusive lexicalization of the recurrent expressions in the text. By excluding the hapax words or those with low frequency (fewer than 4 occurrences) from the process of lexicalization, we automatically exclude the accidental sequences which do not constitute a complex lexeme.

The next stage thus consists of extracting the nominal syntagmas resulting from the validation of the Regular Expressions (RE) that reconstruct the syntactic structures of the most common collocations.

8.4.3 Searching the Syntactic Structures in the Corpus

In this exploratory stage, we verify all the possible REs which will make up the final model of the nominal syntagmas. As will be recalled, several principle syntactic structures for the retrieval of nominal syntagmas were defined in Sect. 7.3.2. During lemmatization, moreover, nouns (N) and adjectives (A) including all ambiguous or non-grammatically classified words were tagged as both A and N in a number of cases.

The syntactical structures in the English language corpus were searched by adapting the combinations defined in Sect. 7.3.2 to English syntax. In particular, this is the position of the adjective with respect to the noun in the structures in which they are combined. Furthermore, the exploratory stage found a number of occurrences of further structures consisting of combinations of adjectives and nouns of type <A + A + N>, <N + N + N>. We then proceeded with the search.

Once all the single REs were validated, a meta-query was created which systematically searches all the subsequent meta-information in the Vocabulary DB CATSEM field, reconstructing the syntactic structures defined earlier.

The complete RE is as follows:

> "CATSEM(N) CATSEM(N)" OR "CATSEM(N) CATSEM(N) CATSEM(N)" OR "CATSEM(A) CATSEM(N)" OR "CATSEM(A) CATSEM(A) CATSEM(N)" OR "CATSEM(N) *of* CATSEM(N)" OR "CATSEM(A) CATSEM(N) *of* CATSEM(N)" OR "CATSEM(N) *of* CATSEM(A) CATSEM(N)" OR "CATSEM(A) CATSEM(N) *of* CATSEM(A) CATSEM(N)".

We can thus recognize 16,326 complex expressions with the syntactic structures mentioned above. By lexicalizing the expressions with at least 5 occurrences, 737 nominal syntagmas are recognized for a total of 7232 occurrences (Table 8.6). Each lexicalized MWE therefore becomes a new vocabulary type.

At this point, the lexical-textual model produces a terminological dictionary obtained directly through the proposed exploratory method.

Thus, the vocabulary passes from 17,119 types (before lexicalization) to 17,831 types. The increase in the number of units is the factor which disambiguates terms and the variety of meanings. It should be noted that the number of entries in the lexicalized vocabulary does not coincide with the sum of the entries before lexicalization and the number of syntagmas because the occurrence of some types has been completely absorbed into the lexicalized syntagmas.

8.5 Analysis of Semantic Dimensions

The following paragraph demonstrates how the selection of recognized MWEs can, in correspondence analysis (CA) (Benzécri 1976, 1992), reconstruct the same categorical structure generated by the entire corpus.

A specific language representation is elaborate on a factorial plan, which graphically presents the combinations of a matrix of [types × categories]. The position of words on the factorial plan is a function of the association of their occurrences in the sub-texts (years), thus expressing their similarity or diversity: two words are close because they are present in the same sub-texts. At the center of the factorial plan are the most common terms among the different languages of the various sub-texts.

Table 8.6 Examples of lexicalized syntagmas (MWE) listed by decreasing total occurrence and distribution in the different journals

MWE	Occurrences (corpus)	Mind	Monist	Philosophical Review	Journal of Philosophy
free will	81	35	2	23	21
theory of knowledge	81	15	7	29	30
history of philosophy	78	4	15	33	26
moral philosophy	72	21	4	21	26
philosophy of science	71	10	6	26	29
philosophy of religion	62	15	10	23	14
human nature	48	7	6	16	19
scientific method	46	7	1	16	22
political philosophy	43	11	6	14	12
philosophy of mind	43	21	1	19	2
theory of truth	42	16	7	3	16
modern philosophy	39	5	2	21	11
personal identity	38	22	5	5	6
common sense	38	12	6	7	13
moral theory	33	12	3	11	7
ontological argument	32	11	5	10	6
critique of pure reason	32	6	1	18	7
social sciences	30	5	0	9	16
ethical theory	30	6	2	13	9
formal logic	30	13	0	6	11
philosophy of history	30	3	1	11	15
natural law	30	8	5	7	10
external world	29	12	0	7	10
philosophy of mathematics	29	4	0	15	10
social psychology	29	5	2	7	15
philosophy of language	29	19	0	6	4
philosophical analysis	27	3	1	12	11
natural science	27	5	3	11	8
moral life	26	3	1	12	10
practical reason	26	12	3	7	4
magic squares	26	0	26	0	0
social philosophy	26	3	2	10	11

Through a correspondence analysis, the row and column elements of the matrix are mathematically formalized as vectors, and the above profiles are represented by points in a multidimensional space. The distances between the lexical profiles are measured using a weighted Euclidean metric (chi-square metric). The complex multidimensional space of the variables is then reduced to a few key factors that can

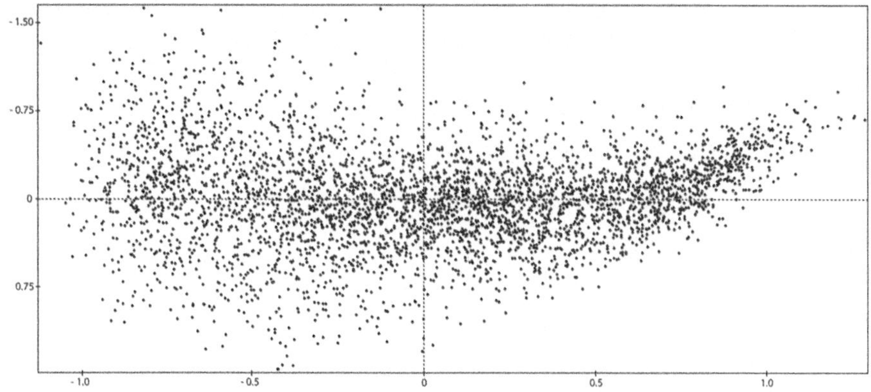

Fig. 8.1 First factorial plane of CA of matrix 1 (4023 words × 141 years). Projection of words

represent, on dimensions named "factorial axes," the relationships between the elements of the data matrix. CA produces the best simultaneous representation of row profiles vs. column profiles in each factorial plan, and on each of its axes (Bolasco 1999).

Two different matrixes were created, whose rows are the types and whose columns are the same variables for Year of Publishing, for each lexical writ with at least 5 occurrences, two different type of classes, specified as follows:

1. The first matrix to be analyzed consists of a contingency table made up of 4023 words of the vocabulary and 141 years resulting from the first tokenization, entirely without lexicalization;
2. The second matrix consists exclusively of 737 lexicalized nominal syntagmas (and 141 years).

The results of the factorial analysis are summarized graphically, allowing us to define the configuration of points on the planes of projection formed by pairs of factorial axes. The distribution of the cloud of the vocabulary types and the categorical structure of the corpus are shown in Figs. 8.1 and 8.2, respectively. This factorial map is the result of correspondence analysis on the first of the two matrixes. As can be seen, the main polarization explains the temporal variable. Moreover, the lexicon shows greater variability on the left side of the factorial plan (Fig. 8.1), while there is a concentration of types on the right side.

Figures 8.3 and 8.4 show the distribution of the examples in the second matrix under analysis, made up of 737 recognized MWE types.

In the different factorial planes shown, we can determine how the structural distribution of the categories (years) remains substantially the same, from the matrix

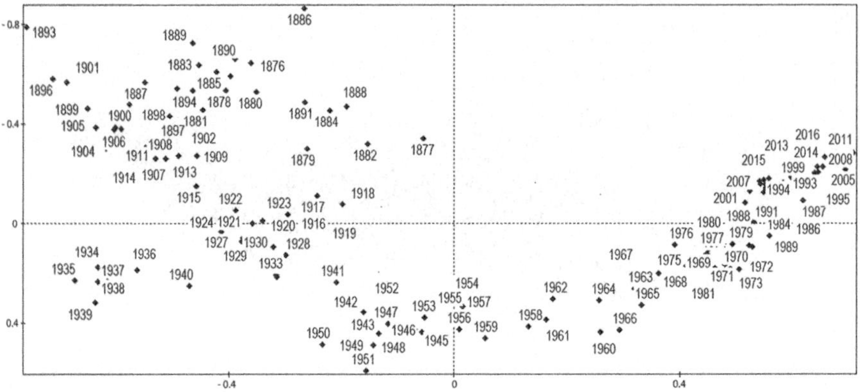

Fig. 8.2 First factorial plane of CA of matrix 1 (4023 words × 141 years). Projection of years

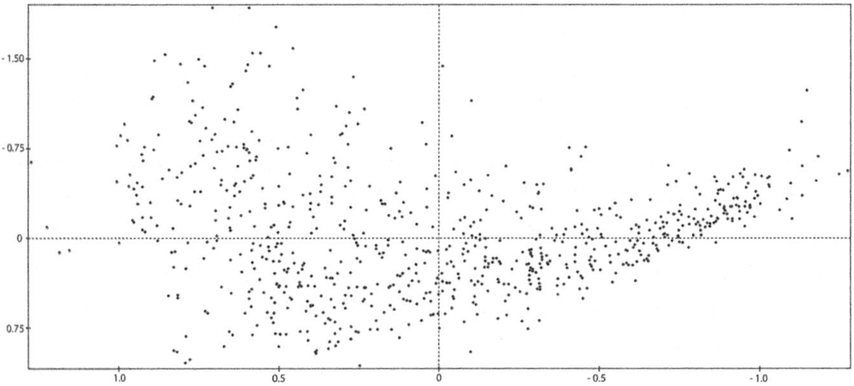

Fig. 8.3 First factorial plane of CA of matrix 2 (737 MWE × 141 years). Projection of MWEs

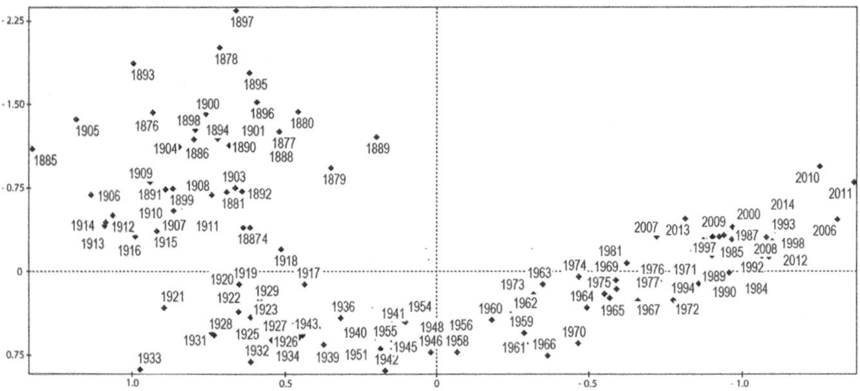

Fig. 8.4 First factorial plane of CA of matrix 2 (737 MWE × 141 years). Projection of years

of the original vocabulary made up 4023 types, to the matrix formed by 737 nominal syntagmas. We can thus emphasize how the terminology which sustains the structure of a vocabulary is presented.

8.6 Conclusions

Syntagmatizing the nominal expressions in the corpus allowed us to disambiguate the terms in the texts, both with respect to the terms and to the meaning of the words. The proposed lexical-textual model, by searching the most common syntactical structures representing nominal syntagmas, allowed us to identify both idiomatic multiwords and their ever-present collocations in specialist texts. Searching syntactic structure was possible thanks to out-of-context tagging, incorporating the grammatical ambiguity resulting from this operation as the main element in the model.

The model is robust in terms of selecting true positives and excluding false positives, in that it is based on the exclusive validation (lexicalization) of the recurring entities found with at least 5 occurrences. A choice of this kind sometimes depends on the objectives involved. Whenever a glossary is wanted for a sector, we proceed with the greatest possible extension of the number of syntagmas. The proposed process allows us to identify the MWEs in the corpus automatically, passing from the words to the terminology with a significant increase in the quality of automatic text analysis. Furthermore, identifying non-ambiguous nominal multiword expressions allows us to overcome the difficulties in accessing the semantic content present in the different text structures. By refining the disambiguation capacity, we provide the information extraction system with the linguistic intelligence needed to filter out unnecessary information and retain useful information.

References

Agirre, E., & Edmonds, P. (2007). *Word sense disambiguation. Text, speech, and language technology*. Dordrecht: Springer.

Benzécri, J.-P. (1976). *L'Analyse des Données. II. L'analyse des correspondances* (2nd ed.). Paris: Dunod.

Benzécri, J.-P. (1992). *Correspondence analysis handbook. Statistics, textbooks and monographs*. New York: Marcel Dekker.

Bolasco, S. (1999). *Analisi multidimensionale dei dati. Metodi, strategie e criteri d'interpretazione*. Roma: Carocci.

Bolasco, S. (2010). *Taltac2.10. Sviluppi, esperienze ed elementi essenziali di analisi automatica dei testi*. Milano: LED.

Bolasco, S. (2013). *L'analisi automatica dei testi. Fare ricerca con il text mining*. Roma: Carocci.

Bolasco, S., & Morrone, A. (1998). La construction d'un lexique fondamental de polyformes selon leur usage. In *JADT*. Nice: Universié de Nice.

Bolasco, S., & Pavone, P. (2010). Automatic dictionary and rule-based systems for extracting information from text. In F. Palumbo & C. N. Lauro (Eds.), *Data analysis and classification. Proceedings of the 6th Conference of the Classification and Data Analysis Group of the Società Italiana di Statistica* (pp. 189–198). Berlin: Springer.

Church, K., & Hanks, P. (1990). Word association norms, mutual information, and lexicography. *Computational Linguistics, 16*(1), 22–29.

Church, K., Gale, W., Hanks, P., & Kindle, D. (1991). Using statistics in lexical analysis. In U. Zernik (Ed.), *Lexical acquisition: Exploiting on-line resources to build a lexicon* (pp. 115–164). Hillsdale: Lawrence Erlbaum Associates.

De Mauro, T. (1999–2007). *Grande Dizionario Italiano dell'Uso (GRADIT)*. Torino: Utet.

Dunning, T. (1993). Accurate methods for the statistics of surprise and coincidence. *Computational Linguistics, 19*(1), 61–74.

Elia, A. (1995). Per una disambiguazione semi-automatica di sintagmi composti: i dizionari lessico-grammaticali. In S. Bolasco & R. Cipriani (Eds.), *Ricerca Qualitativa e Computer* (pp. 112–141). Milano: Franco Angeli.

Elia, A. (1996). Per filo e per segno: la struttura degli avverbi composti. In E. D'Agostino (Ed.), *Sintassi e Semantica* (pp. 167–263). Napoli: ESI.

Grigolli, S., Maltese, G., & Mancini, F. (1995). Un prototipo di lemmatizzatore automatico per la lingua italiana. In S. Bolasco & R. Cipriani (Eds.), *Ricerca Qualitativa e Computer* (pp. 142–155). Milano: Franco Angeli.

Justeson, J. S., & Katz, S. M. (1995). Technical terminology: Some linguistic properties and an algorithm for identification in text. *Natural Language Engineering, 1*, 9–27.

Lenci, A., Montemagni, S., & Pirrelli, V. (2005). *Testo e Computer. Elementi di linguistica computazionale*. Roma: Carocci.

Morrone, A. (1993). Alcuni criteri di valutazione della significatività dei segmenti ripetuti. In *Actes des secondes Journées Internationales d'Analyse Statistique de Données Textuelles* (pp. 445–453). Paris: Anastex S. J.

Pavone, P. (2010). Sintagmazione del testo: una scelta per disambiguare la terminologia e ridurre le variabili di un'analisi del contenuto di un corpus. In S. Bolasco, I. Chiari, & L. Giuliano (Eds.), *Jadt 2010—Statistical analysis of textual data* (Vol. 1, pp. 131–140). Roma: LED.

Rouget, C. (2000). *Distribution et sémantique des construcuions Nom de Nom*. Paris: Honoré Champion.

Sag, I. A., Baldwin, T., Bond, F., Copestake, A., & Flickinger, D. (2002, February). Multiword expressions: A pain in the neck for NLP. In *International Conference on Intelligent Text Processing and Computational Linguistics* (pp. 1–15). Heidelberg, Berlin: Springer.

Salem, A. (1987). *Pratique des segments répétés. Essai de statistiques textuelle*. Paris: Klincksieck.

Schmid, H. (1994). Probabilistic part-of-speech tagging using decision trees in proceedings of international conference on new methods in language processing. In *Proceedings of International Conference on New Methods in Language Processing*. Manchester.

Schmid, H. (1995). Improvements in part-of-speech tagging with an application to German. In *Proceedings of the ACL SIGDAT-Workshop*. Dublin.

Sinclair, J. (1991). *Corpus concordance collocation*. Oxford: Oxford University Press.

Chapter 9
Functional Data Analysis and Knowledge-Based Systems

Matilde Trevisani

Contents

9.1 Introduction 168
9.1.1 An Outline of the Method 169
9.2 Related Literature 170
9.3 Method in Detail 173
9.3.1 Normalisation of Word Trajectories 173
9.3.2 Word Trajectory Filtering 175
9.3.3 Curve Clustering 176
9.3.4 Substantive Expertise 178
9.4 Illustration 178
9.4.1 Corpus Collection and Construction 178
9.4.2 Normalisation 179
9.4.3 Filtering 180
9.4.4 Curve Clustering 182
9.4.5 Substantive Expertise 184
9.5 Concluding Remarks 185
References 186

Abstract In the present study, the challenge is whether a distant reading of the history of a discipline can be achieved by analysing the temporal evolution of keywords retrieved from papers in the discipline's mainstream journals. This calls for the so-called knowledge-based system (KBS), i.e. a computer-based system that supports human learning not only by acquiring and manipulating large volumes of data and information, but also by integrating knowledge from different sources. In this chapter, we introduce a KBS that, starting from a large database of texts retrieved from scientific articles published over a lengthy period by a selection of the discipline's premier journals, leads to the construction of a well-founded corpus of scientific literature and from this to a possible outline of the discipline's history. Our work is based on the idea that the temporal course of a word occurrence is a

M. Trevisani (✉)
University of Trieste, Trieste, Italy
e-mail: matildet@deams.units.it

A. Tuzzi (ed.), *Tracing the Life Cycle of Ideas in the Humanities and Social Sciences*, Quantitative Methods in the Humanities and Social Sciences,
https://doi.org/10.1007/978-3-319-97064-6_9

proxy of the word's life cycle. We then adopt a functional data analysis (FDA) approach under which we first reconstruct words' life cycles. Second, by clustering words with similar life cycles, we detect any prototypical or exemplary temporal patterns representing the latent dynamics of word micro-histories. The major dynamics uncovered at this stage are then submitted to subject matter experts for interpretation and guidance in decision-making, thus making it possible to trace a history of the discipline. Moreover, we propose several kinds of data normalisation which involve different concepts of life cycle similarity and hence a different reading of the history of the discipline under examination.

Keywords Distant reading · Trajectory normalisation · Curve clustering · Clustering validation · Clustering agreement

9.1 Introduction

Scientometrics studies the evolution of science quantitatively, by analysing publications. One of its main objectives is to develop information systems that can help explore the enormous number of scientific articles that are constantly being published. In the present study, the challenge is whether a "distant" reading of the history of a discipline (or, more generally, a field of knowledge) can be achieved by analysing the temporal evolution of keywords retrieved from papers in the discipline's mainstream journals. This calls for a so-called knowledge-based system (KBS), i.e. a computer-based system that supports human learning not only by acquiring and manipulating large volumes of data and information, but also by integrating knowledge from different sources. In this chapter, we introduce a KBS that, starting from a large database of texts retrieved from scientific articles published over a lengthy period by a selection of premier journals in a specific discipline, leads to the construction of a well-founded corpus of scientific literature (organised as a "keywords × time points" matrix) and from this to a possible outline of the discipline's history. Corpus creation is assisted by knowledge from linguistics experts captured in the system's knowledge base, while knowledge from experts in the specific domain being investigated assists the learning process in both interpretation and decision-making, potentially enabling it to culminate in a conclusive reading (or readings) of the history.

The underlying assumption of the research project presented in this book is that the temporal evolution of words (in terms of their occurrence) reflects the relevance of the corresponding concepts (ideas, themes, research problems) in the scientific discourse over time. In making this assumption, we are aware that the timeframes for methods and research fields yielded by our analysis reflect the moment when they became established in the scientific community and spread to the literature (which is necessarily later than the time these methods are introduced or interest is first expressed in these fields).

Our work is based on the idea that the temporal course of a word occurrence is a proxy of a word's diffusion and vitality, i.e. of the word's life cycle (Trevisani and Tuzzi 2012, 2013a, b, 2015, 2018; Tuzzi and Köhler 2015). We then adopt a functional data analysis (FDA) approach where the observations (occurrences) through time are viewed as a realisation of an underlying continuous function representing the temporal development of a word.

In the FDA approach, we first reconstruct words' life cycles. Second, by clustering words with similar life cycles, we detect any prototypical or exemplary temporal patterns representing the latent dynamics of word micro-histories. Examples of prototypical patterns include essentially increasing, decreasing or constant trends, trends with an isolated peak for briefly faddish words, or roughly bell-shaped trends for words which had a golden age and then disappeared, etc. The major dynamics uncovered at this stage are then submitted to subject matter experts for interpretation and guidance in decision-making, thus making it possible to trace a history of the discipline in question.

Intrinsically connected to the primary aim of this study is what type of information the time trajectory of a word occurrence should contain in order to correctly construct and compare words' life cycles and discover the important dynamics of the ideas latent in word groups. Should timing or synchrony be the sole determinant when assessing the similarity of words' life cycles, or should word popularity be considered significant when comparing different words? For FDA, in other words, should we compare word curves only on the basis of their phase variation, or also by accounting for their amplitude variability? This question arises from a typical feature of textual data, the so-called Large Number of Rare Events (LNRE) property, i.e. the presence of a large number of word-types whose probability of occurring is quite low, which implies data sparsity and high skewness. Regardless of any decision concerning these issues, preliminary data processing is advisable in order to adjust for the uneven size of subcorpora (number of texts and their size in word-tokens) over time and hence regularize the "signal". Further data transformation depends on the study's specific aims and is crucial for a consistent reading of results. In this chapter, we propose several kinds of data normalisation which involve different concepts of life cycle similarity and hence a different reading of the history of the discipline under examination.

9.1.1 An Outline of the Method

Given a knowledge field of interest, the KBS consists of two main stages:

1. An information retrieval process that, starting from a large database of scientific articles published by a selection of premier journals in the field, leads to the creation of a well-founded corpus of scientific literature.

2. A statistical learning process that leads to the reconstruction of the important dynamics underlying word micro-histories and hence to an outline of the overall evolution of the knowledge field in four steps:
   (a) Normalisation of time trajectories of word (raw) frequencies, chosen according to aspects of life cycles that are considered substantive when comparing words.
   (b) Filtering time trajectories of word (normalised) frequencies, interpreted as functional data (FD) and thus represented as smooth functions.
   (c) Curve clustering (CC) to detect all important dynamics underlying the evolution of groups of word micro-histories.
   (d) Interpretation by expert opinion to decipher detected dynamics and thus compose a narrative of the evolution of the knowledge field as a whole.

We adopt a basis function approach to filtering with a B-spline basis system. Moreover, we take a distance-based approach to CC and use a k-means algorithm for FD combined with an appropriate metric for measuring distance between curves. In the illustration, we use the Euclidean distance. While interpreting, experts can formulate new research questions that may lead to further insights. If CC yields concurrent solutions, the experts can decide on one or more historical narratives for the knowledge field in the period examined.

We situate our methodological choices in the literature in Sect. 9.2, and describe the method in greater detail in Sect. 9.3.

9.2 Related Literature

The objective we pursue here has some analogies with other research areas though the approaches they propose are markedly different and cannot answer our particular question effectively. We list three major lines of research which are alternative to ours.

Quantitative linguistics often deals with textual data consisting of temporal sequences of linguistic units and, generally, addresses the problem of reading the evolution of a linguistic phenomenon over time by applying linguistic laws, Fourier analysis (and similar methods) or time series analysis. In our study, however, a word trajectory is very unlikely to show a regular behaviour (e.g. that fits a function) and is only apparently a matter of time series analysis. The latter focuses on studying the correlation of observations over time and, normally, seeks a model for prediction. By contrast, the word life cycle is the primary, indivisible unit of our analysis (the functional datum) and our primary goal is to recognise temporal shapes or curves from raw word trajectories.

Topic modelling shares similar aims with our perspective, but only to a certain extent. When topic modelling is applied to documents referenced to time points, co-occurrence (of words within documents) analysis—on which topic modelling is based—can be transmuted to our analysis (Griffiths and Steyvers 2004).

Nevertheless, differences from our approach are evident from their primary aim: unveiling topics (hence mapping science and tracking its evolution) versus tracing life cycles of words (hence dynamics of temporally homogeneous bundles of words in order to decipher the history of a knowledge field). Topic modelling produces clusters of words that should reflect a topic when they appear together in documents (but the shape of word trajectories is not relevant), whereas our approach leads to clusters of words that should evolve similarly over time (but that might represent different topics, different approaches, or different schools of thought). Additionally, it has been shown that Latent Dirichlet Allocation (LDA)—the standard method used for topic modelling—is not the best approach for analysing corpora that include texts of limited length (e.g. titles of articles, Trevisani and Tuzzi 2018 and references therein).

Scientometrics (to which topic modelling connects) or, more in general, quantitative methods for mapping knowledge domains from scientific article databases, are based on term and/or citation co-occurrences in documents, possibly observed over time in order to reconstruct a field's evolution. Recently, many researchers have adopted generative probabilistic models for topic detection and tracking (TDT) or, in general, dynamic science mapping. These models include LDA (despite certain shortcomings that undermine its role—viz., it requires that the number of topics be specified in advance and tends to an even distribution of topics—if the focus is on finding emerging topics and how they evolve over time) and the hierarchical Dirichlet process (HDP), a nonparametric Bayesian model which can automatically decide the number of topics, and is thus considered more competent than LDA in dynamic topic analysis (Ding and Chen 2014). However, traditional topic analysis approaches are relatively static, as they ignore any changes (in both the external representation and the internal content of a scientific topic) that may occur over time. Two recent studies dealing with topic changes and emerging topic detection (ETD) are Zhang et al. (2016, 2017). Both use a term clumping process for core term retrieval, after which the first applies k-means-based clustering to obtain topics and finally produces a "roadmapping" that blends historical analysis and expert-based forecasting. The second applies an LDA-based topic model to profile the topic landscape, then a model of scientific evolutionary pathways to detect topic changes and to indicate emerging topics, and lastly, a prediction model to foresee possible topic trends. Another approach to analysing the thematic evolution of a given research field is presented in Cobo et al. (2011) and has been incorporated in SciMAT (Cobo et al. 2012). Recently, the traditional topic evolution map based on text corpora has been extended to more complex subjects like cross-media data (Zhou et al. 2017) and memes (Shabunina and Pasi 2018).

In conclusion, science mapping research is based on co-occurrences in documents possibly observed over time, while our work considers term co-occurrence solely in time, as our primary focus is the temporal evolution of terms. More importantly, our approach differs conceptually from the main alternatives that address the problem of knowledge evolution, such as those developed for TDT, ETD and, generally, for dynamic knowledge mapping in scientometric studies. Our analysis focuses on detecting important dynamics each of which represents the temporal evolution of a group of words. Thus, on principle, different themes, research fields and approaches can be represented within the same group of words. Conversely, "topic-centered" methods focus first on the structure of science and detecting topics, and then on tracking their evolution. As a consequence, words that represent the same topic may have an irreconcilable temporal evolution. Moreover, topic evolution can only be a roadmap, i.e. an abstract description (the average evolution of words grouped by co-occurrence) of basic movements over time. Additionally, the abstract definition of topics is subjected to continuous destruction and reconstruction by time, making topic tracking a fragile and questionable artefact.

Finally, our choice of specific statistical tools is underpinned by the literature as follows.

The basis function approach is the most widely used for representing FD, and B-splines are a very flexible basis system for non-periodic FD (Ramsay and Silverman 2005). Moreover, B-splines enable us to recognise continuous and regular curves, and hence more easily interpretable shapes. Other systems, e.g. wavelets, can be better suited to the typical bumpy trend of word trajectories (Trevisani and Tuzzi 2015). Upstream, we decided for a distance-based approach, as one of our objectives was to set up an exploratory and mostly automated procedure. In fact, the procedure is called upon to look for interesting patterns—without prescribing any specific interpretation—to be submitted to experts who can potentially formulate new hypotheses and research questions. This eminently exploratory task requires the procedure to be fast and relatively easy to use and understand even by non-statisticians in interdisciplinary groups involved in research projects. The alternative or model-based approach is typically chosen for confirmatory analyses and is generally more demanding in terms of computing and inferential expertise. In a previous study (Trevisani and Tuzzi 2015), we used a functional mixed (normal mixture) model based on a wavelet-based decomposition which proved effective in accommodating the irregularity of word curves and the high inter-word variability, as well as being computationally efficient in a modelling context with high-dimensional data.

Once opted for distance-based methods, *k*-means type clustering algorithms have been widely applied to FD, especially when combined with the finite basis expansion approach. Other strategies which extend the classical *k*-means algorithm with FD are essentially based on functional principal components. However, they are recent extensions, rarely used and, thus, less justifiable as the basis for our explorative approach (some interesting overviews of strategies for clustering FD are provided by Jacques and Preda 2014, and Wang et al. 2016).

Lastly, we opted for the Euclidean distance (L_2 metric) for measuring distance between curves since conventional distances between raw data evaluating a one-to-one mapping of each pair of sequences meet our needs. In fact, one of our objectives is to compare curve profiles after data transformation. Accordingly, our strategy entails first transforming data and then seeing what this involves for clustering results by using a distance measure that can approximate the area between two curves as simply as possible. In our application, we used L_2 as it is the most popular metric though an equally simple alternative would be the L_1 or Manhattan distance.

The alternative way of directly choosing a dissimilarity measure which is invariant to specific distortions of the data is not suitable here, as filtering is to be performed on preprocessed data.

9.3 Method in Detail

The first stage of the KBS consists in compiling and constructing the corpus, as Chap. 6 describes in detail. Here, we will review only the main steps of the information retrieval procedure.

Corpus compilation involves a preliminary selection of data sources, i.e. choosing outstanding journals that can cover main topics and represent the temporal evolution of the knowledge field. Text harvesting follows, i.e. downloading information (all references, numbers, issues, volumes) from journal archives to make up the article database. At the end of this step, a diachronic corpus, i.e. a collection of texts including information on their time period, is created. Text under consideration consists of titles and/or abstracts and/or full texts of the articles. Moreover, a corpus is typically organised into subcorpora, or groups of texts sharing the same time reference, thus generating a sequence of text sets along a chronological sequence of time points.

Corpus construction and pre-processing (Chap. 7) involve identifying all words (in the tokenisation stage, words are sequences of letters isolated by means of separators), as well as other possible forms of tagging: stemming, or transforming words into stems; identifying (and ranking) stem-segments (or n-stem-grams, i.e. sequences of stems); tagging keywords, or identifying all words (stems and stem-segments) relevant to the specific knowledge field (e.g. by matching the corpus vocabulary with item lists of relevant glossaries for the knowledge field); and thresholding, or selecting all keywords with frequencies at least equal to an appropriate threshold. Finally, the corpus is represented by a words × documents/time points contingency table containing the frequencies of the selected keywords (by row) along the time points (by column) of the period considered.

The second stage of the KBS consists of a stepwise process of statistical learning that enables a distant reading of the diachronic corpus.

9.3.1 *Normalisation of Word Trajectories*

A diachronic corpus is typically characterised by the following features.

1. Size of subcorpora (number of texts and their size in word-tokens) may vary greatly over time.
2. The LNRE property of textual data, i.e. a large number of word-types whose probability of occurring is quite low, which implies:

(a) The total frequency (or popularity) of individual words in the entire corpus is highly variable,
(b) The frequency spectrum by time point is highly asymmetric,
(c) Frequency sparsity, i.e. many cells of the contingency table have small counts or are empty.

In Sect. 9.4, features (1) and (2) are illustrated in Figs. 9.1 (subcorpora size) and 9.2 (original word trajectories), respectively.

As the foregoing considerations indicate, raw frequency normalisation is necessary to reconstruct and compare the temporal evolution of words. In particular, a form of normalisation by time point should be regarded as preliminary in order to adjust the uneven size of subcorpora across time and hence regularise the "signal". A further form of normalisation by word might be appropriate in order to adjust the great disparity in word popularity, thus making it possible to compare word trajectories by timing (synchrony) regardless of height (popularity). We envision several types of normalisation, of which Table 9.1 gives an excerpt.

Normalisation by column can be obtained, for example, from dividing raw frequencies by the number of texts (option c_1) or the total number of word-tokens in texts (c_2) for each subcorpus (/ time point), still, by the column sum (c_3) or the column maximum frequency (c_4) of the data table. Normalisation by row can be obtained, for example, from dividing the raw frequencies by the row sum (r_1) or the row maximum frequency (r_3) of data table, still, by computing the z-scores of word raw frequencies (r_2). Double (by both row and column) normalisation (*d*) serves to fix both (1) and (2). The calculation of specific double normalisations (d_1 and d_2) is illustrated in Sect. 9.4 (Figs. 9.3 and 9.4).

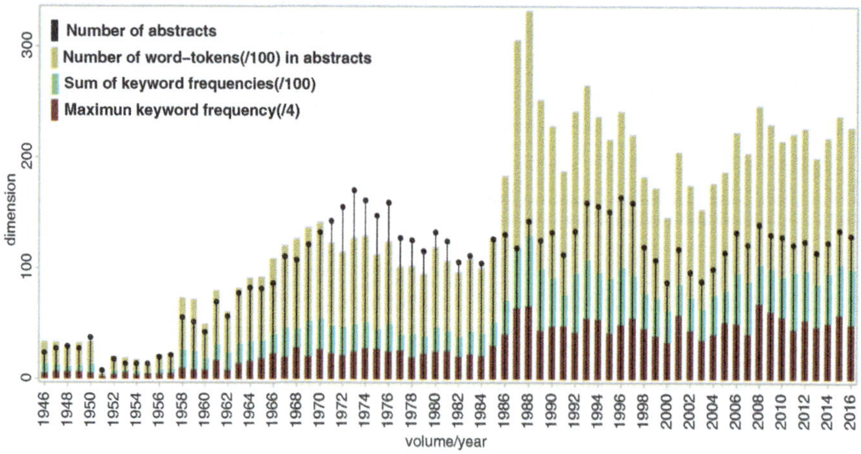

Fig. 9.1 Subcorpora size: for each volume, number of abstracts (dot-line), total number of word-tokens in abstracts/100, sum of keyword frequencies in data table/100, maximum keyword frequency in data table/4

Table 9.1 Excerpt of normalisation plan

Normalisation:	By col	Subcorpus		Words × documents table		
By row		# texts	#tokens	col sum ($\sqrt{\cdot}$)	col max freq	
Row sum		d	d	d_1	d	r_1
z-score by row		d	d	d	d	r_2
Row max freq		d	d_2	d	d	r_3
		c_1	c_2	c_3	c_4	

9.3.2 Word Trajectory Filtering

From an FDA perspective, the functional observation $\mathbf{y}_i = \{y_{ij}\}$ of word i consisting of the set of (normalised) frequencies at time points $t_j = t_1, \ldots, t_T$, for each $i = 1, \ldots, N$, is viewed as a realisation of an underlying continuous function $x_i(t)$—sufficiently smooth or regular—representing the word's temporal evolution. As $\mathbf{y}_i$ is a noisy observation of the underlying $x_i(t)$, an adequate model of their relationship is $\mathbf{y}_i = x_i(\mathbf{t}) + \boldsymbol{\varepsilon}_i$, where $\mathbf{t} = \{t_j\}$ and $\boldsymbol{\varepsilon}_i = \{\varepsilon_{ij}\}$ is a zero mean vector with dispersion matrix $\mathrm{Var}(\boldsymbol{\varepsilon}_i) = \Sigma_\varepsilon$. In the standard model, the ε_{ij}s, often termed "measurement errors", are assumed independent across j and homoscedastic with $\sigma_{ij}^2 = \sigma^2$, but, in a more general case, Σ_ε can be regarded as full and time dependent. The following choices are adopted for filtering $x_i(t)$ from $\mathbf{y}_i$ (see Trevisani and Tuzzi 2018, for a detailed description and rationale).

We adopt the basis function approach for representing FD as smooth functions where $x_i(t)$ is expressed as a finite linear combination

$$x_i(t) = \sum_{k=1}^{K} c_{ik}\phi_k(t) \quad c_{ik} \in \Re, K < \infty$$

of real-valued functions φ_k called basis functions (Ramsay and Silverman 2005).

We consider B-spline bases, the most popular basis system for building spline functions, which are piecewise polynomials joined smoothly at the interior nodes.

As regards the positioning of knots—the values of t at which adjacent segments are joined, a direct and reasonable choice is that of placing knots at each point of observation t_j.

We adopt the roughness penalty or regularisation approach for smoothing FD, whereby the estimate of x_i is the function minimising the penalised residual sum of squares $\mathrm{PENSSE}(x_i) = \mathrm{SSE}(x_i) + \lambda\, \mathrm{PEN}_r(x_i)$ where $\mathrm{SSE}(x_i)$ is the residual sum of squares measuring the fit to the data, $\mathrm{PEN}_r(x)$ is the penalty term measuring a function roughness (by the integrated squared rth derivative over the observation time, $\mathrm{PEN}_r(x) = \int [D^r x(s)]^2 ds$) and λ is a smoothing parameter. Thus, λ measures the tradeoff between fit to the data and roughness of the function x: as $\lambda \to 0$ the fitted curve approaches an interpolant to the data, as $\lambda \to \infty$ the condition $\mathrm{PEN}_r(x) \to 0$ means the fitted curve is a spline of order r.

Choosing λ is part of the model selection issue. A standard practice for choosing λ is to use cross-validation (CV). When tuning a smoothing parameter, a common choice is the leave-one-out CV, however, it may be computationally intensive especially for large sample sizes and lead to under-smoothing. Generalised cross-validation (GCV), $\mathrm{GCV}(\lambda) = T / (T - \mathrm{df}(\lambda))^2\, \mathrm{SSE}(\hat{x}_i)$, provides a convenient approximation to leave-one-out CV for linear fitting under squared error loss. df(λ) is the effective degrees of freedom under regularisation, which is monotone decreasing in λ with maximum equal to K when $\lambda = 0$. GCV can sufficiently remedy the tendency to under-smoothing unless the sample size is small or moderate (Lukas et al. 2016; Ramsay and Silverman 2005).

We smooth the data by varying

- Spline order m (from 1 to 8);
- Roughness penalty order r: Besides the standard $r = m - 2$, $r = 2$ for $m > 3$, $r = 1$ for $m > 2$, $r = 0$:
- Smoothing parameter λ over an appropriate range of values ($\log_{10}\lambda$ from −6 to 9).

The GCV criterion is used to select the optimal smoothing.

Calculation is carried out in the R software environment using the *fda* library and an ad hoc routine (R core team 2017). Optimal smoothing selection is illustrated for the case of d_2 normalised data in Sect. 9.4 (Figs. 9.5 and 9.6).

9.3.3 Curve Clustering

In a distance-based approach to CC, we apply a k-means algorithm for FD where the distance between curves is approximated by using the discretely observed evaluation points of the estimated curves $x_i(t)$ (Jacques and Preda 2014). We use L_2 metric, though several options for distance besides the conventional ones can be taken from the broad range of dissimilarity measures available for time series clustering (Montero and Vilar 2014). Moreover, for each cluster number (k from 2 to an appropriate range maximum), we re-run the algorithm starting from 20 different initial configurations set through the k-means++ seeding method.

At this step of our KBS, clustering validation is performed by using the internal information of the clustering process to evaluate the goodness of a clustering structure without reference to external information. External validation is postponed to the next step and consists of an informal assessment by subject matter experts who can decide to what extent a clustering is meaningful to them (see subsection below). Internal validation can be also used to decide what the most appropriate number of clusters is in a certain application. In our research context, no "natural" or "true" clusters exist in the available data, so there is not even a "true" number of clusters. Since we only have to be reasonably confident that nothing important is left unexplored, the idea is that of identifying a set of best candidates for cluster number by pooling the ratings from a large number of clustering quality criteria (about 50, see Desgraupes 2016, and Genolini et al. 2015). It is well known that one index does not fit all situations, rather, the many existing indexes can be grouped into different types each measuring a different aspect of clustering quality. Accordingly, and given the exploratory and evocative task of our clustering, we have gathered a large basket of indexes in order not to favour any single criterion, as each in principle is equally valid. These include measures of within-cluster homogeneity, e.g. *Ball-Hall, Banfeld-Raftery, C-index, Gap, Krzanowski-Lai, Marriot, Scott-Symons*; of between-cluster separation, e.g. *Rubin, Scott, Ratkowsky-Lance*; and of their combination, e.g. *Calinski-Harabasz, Davies-Bouldin, Dunn* and its generalisations, *Gamma, Hartigan, McClain, PBM, Point-Biserial, Ray-Turi, SD, Silhouette, Friedman, Xie-Beni, Tau*; as well as measures of similarity between the empirical within-cluster distribution and distributional shapes such as the Gaussian distribution, e.g. *BIC, AIC* and their variants.

In detail, cluster number selection includes the following steps, in order:

- A cluster number ranking is computed for each quality index.
- All the rankings are pooled and, for each cluster number, the frequency of being ranked first (top-1), second (top-2), third (top-3) and fourth (top-4) is calculated.
- An ordered set of best candidates for cluster number is retrieved from a qualitative inspection of the graphical representation of the frequencies of being in the top four positions for each cluster number (an R code that essentially mimics the visual selection was developed in order to make the procedure automated without the need for human "eye").

An example of this procedure is illustrated in Sect. 9.4 (Fig. 9.7).

For each candidate for cluster number, the best partition between the 20 replications must be chosen. But, in our approach to cluster number selection, there may be multiple criteria that ranked the candidate in the top positions. Then, we compare, for each cluster number, all the distinct partitions resulting from these multiple criteria by concordance measures. In particular, we consider the Rand index (Rand 1971) as measure of agreement between two clusterings and propose a generalisation of it for comparing more than two clusterings, thus obtaining a measure of concordance between multiple clusterings. Moreover, this "multiple" Rand index can be computed at several levels (of individual words, of single clusters as well as of the overall partition), thus offering a measure of stability of clustering results for each of these levels. The standard Rand index calculates the rate of pairs of units that are classified in the same way (i.e. pairs that are in the same cluster or in different clusters, respectively) in both clusterings. We use the standard version for choosing the best partition for each cluster number as the one that maximises the average of the agreement measures with each of the other partitions selected for the cluster number. Namely, the best partition for each cluster number is the one that best mediates between all the partitions of the cluster number. In addition, we use the multiple Rand index to provide a measure of individual agreement for each word and thus investigate whether a particular word is consistently grouped or separated from other words by different partitions. The information on individual words can help to screen out “wird” words with very low agreement measures. The average of such measures of individual agreement over a cluster gives a measure of agreement per cluster; the average over the entire corpus coincides with the multiple Rand index of global agreement between multiple clusterings here proposed.

The R software environment contains several *k*-means implementations as well as libraries for computing clustering quality criteria. Our procedure uses the kml routine (Genolini et al. 2015) which is designed specifically for longitudinal data and provides various efficient methods of *k*-means initialisation. The *clusterCrit*, *cclust*, *clusterSim* and *kml* libraries are used to source the quality criteria considered by our method. Ad hoc functions have also been developed for specific criteria and for calculating the multiple Rand index of local and global agreement between partitions.

9.3.4 Substantive Expertise

Clustering results obtained with the cluster numbers selected as the best candidates are then presented to subject matter experts. To facilitate the comparison between different groupings (partitions with different number and composition of groups), we assess the congruence of different partitions by calculating their indexes of agreement (Wagner and Wagner 2007) and visualising set overlaps (see mosaic plots in Chap. 6). Experts try to interpret the latent content of word groups as a consistent ensemble of topics, methods and research areas, and to identify temporal phases and processes from group dynamics in order to reconstruct a historical narrative of the knowledge field. Where possible, they will be instrumental in suggesting other analyses.

9.4 Illustration

For illustration, we apply the KBS to the corpus of abstracts of scientific papers published by the *Journal of the American Statistical Association* (JASA) in the time span 1946–2016 in order to trace a history of statistics. The extensive analysis is presented in Chap. 6. Here, we will track the main steps to provide an exemplification of the theoretical method outlined above.

9.4.1 Corpus Collection and Construction

JASA is the oldest statistical journal and has long been considered the world’s premier review in its field. We downloaded from online resources all references, abstracts and metadata of articles published in the period 1946–2016 (71 years, from Volume No. 41, Issue No. 234, to Volume No. 111, Issue No. 516). Abstracts of articles constitute the text corpus considered in this study. The corpus includes 7221 abstracts, 1,029,251 word-tokens (word occurrences) and 26,686 word-types (distinct words). After stemming, all potentially relevant stem-segments are identified. Relevant statistical keywords (stemmed words and sequences of stemmed words) are then tagged.

Lastly, fixing the threshold at 50, 1351 keywords are selected. At the end, the corpus yields a 1351 (words) × 71 (time points/volumes) contingency table (see an excerpt in Table 6.2, Chap. 6).

9.4.2 Normalisation

For illustration, we choose to transform data (Fig. 9.2) by the double normalisations d_1 and d_2 (Table 9.1). Let n_{ij} be the raw frequency of word i at time point/volume j, $n_{i.}$ the i-row sum, $n_{.j}$ the j-column sum and n the matrix total of the corpus table. Then, the d_1 normalised frequency is computed as $y_{ij} = n_{ij}/(n_{i.}\sqrt{n_{.j}/n})$ and is equivalent to calculating a χ^2 distance between original word profiles if the Euclidean distance is used as measure of dissimilarity ($n_{.j}/n$ is the j-column mass in correspondence analysis). Note that this double normalisation produces a somewhat

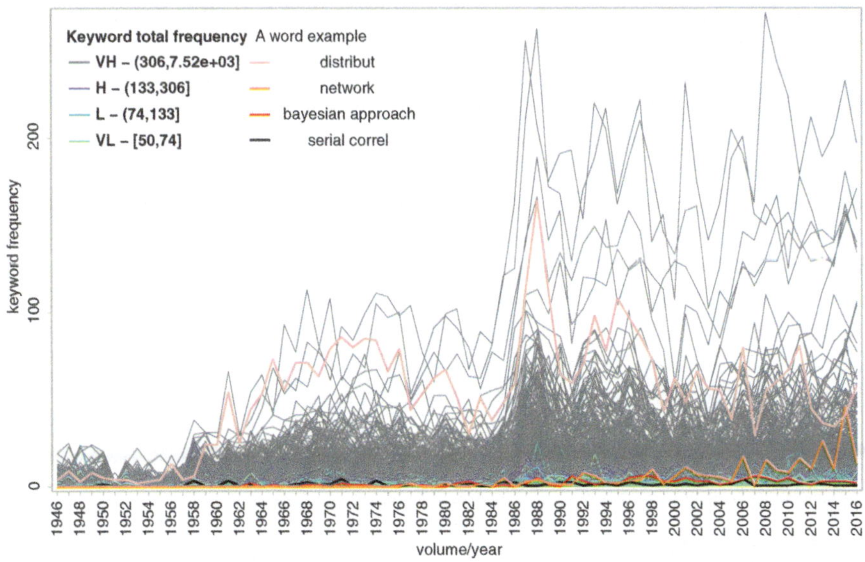

Fig. 9.2 Word trajectories (original data): y-axis represents the raw word frequency for each volume; x-axis represents the volume publication year; line colour identifies the word frequency class (Very Low, Low, High and Very High denote equal-frequency intervals of total word frequency). A word example for each class is superimposed

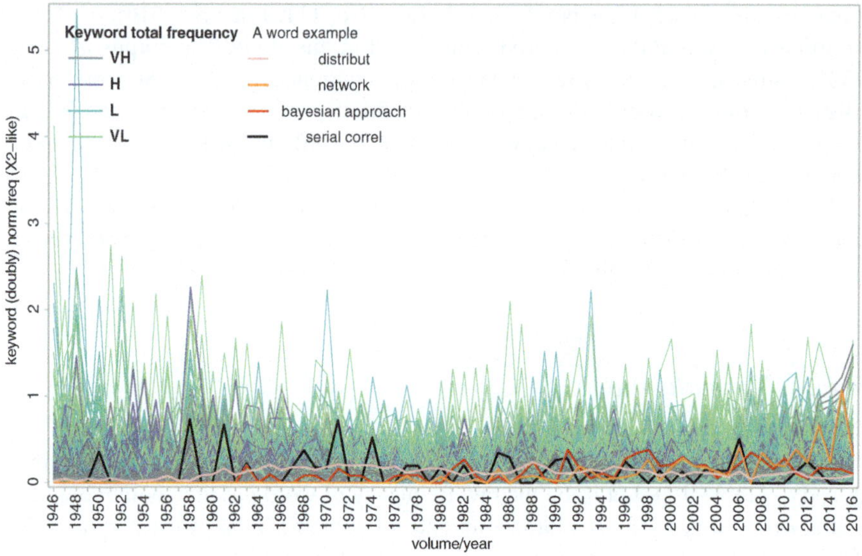

Fig. 9.3 Keyword trajectories (doubly normalised data, d_1 or χ^2-like)

reversed asymmetry (low-frequency words tend to dominate being the associated curves larger in amplitude; see Fig. 9.3). This is mainly due to a greater sparsity of low-frequency words across time. The problem of asymmetry is instead substantially reduced by d_2, thus allowing a comparison between curves mainly in terms of horizontal or phase variation (Fig. 9.4). Let N_j be the total number of word-tokens in the subcorpus j and M_i the i-row maximum frequency of the column-normalised frequencies n_{ij}/N_j. Then, the d_2 normalised frequency is computed as $y_{ij} = n_{ij}/(M_i N_j)$. However, d_2 cannot completely remedy the problem of sparsity. Rare words tend to have sparse trajectories (i.e. to have zero or almost zero frequency for relatively long stretches of the period, either continuous or intermittent) and very high differences of amplitude along the trajectory (being frequency values little spaced; see, e.g. panels of *semiparametric model* and *realistic*—at left-most, third and sixth rows, respectively, in Fig. 6.15, Chap. 6). On the contrary, words with a high or very high popularity tend to have non-negligible frequencies for most of the period and trajectories with lower differences of height (being the grid of frequency values finer; see, e.g. panels of *nonparametric* and *simulation*, left-most panels of first row in Fig. 6.15).

9.4.3 Filtering

Optimal smoothing for d_2 normalised data is achieved with spline order $m = 7$ and smoothing parameter $\lambda = 10^{1.5}$ (df = 6.56) under a roughness penalty of order $r = 1$ (Fig. 9.5).

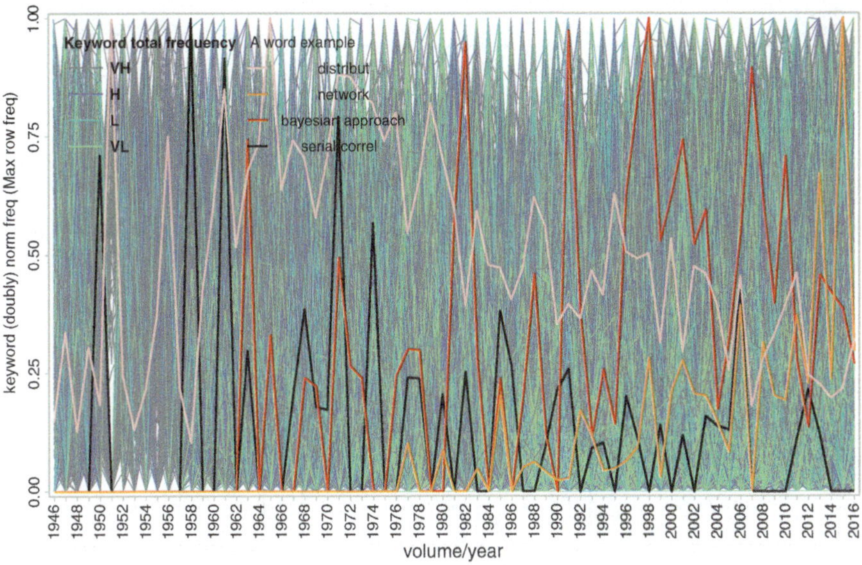

Fig. 9.4 Keyword trajectories (doubly normalised data, d_2)

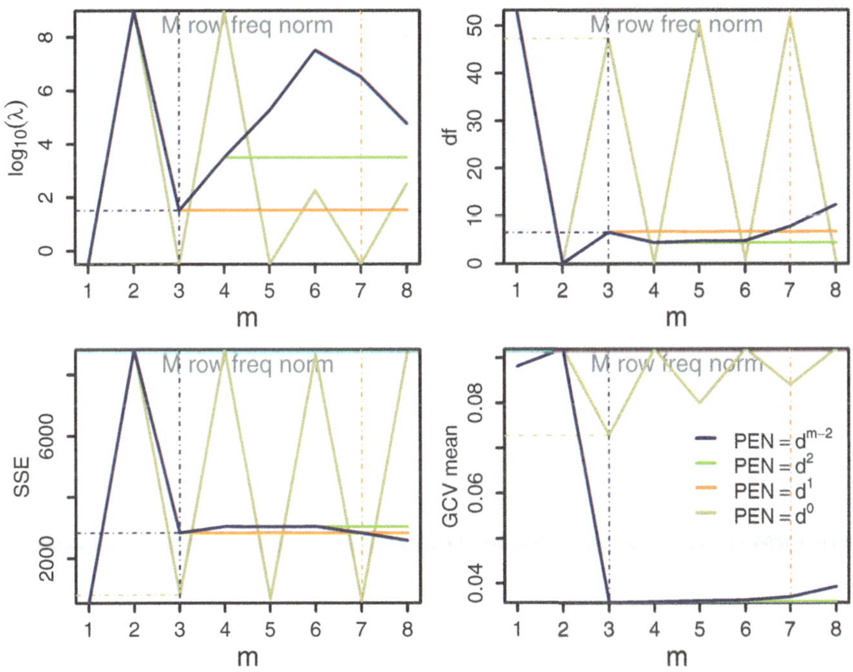

Fig. 9.5 Smoothing selection: optimal λ (top-left) and corresponding effective degrees of freedom (df, top-right), sum of square errors (SSE, bottom-left) and GCV (bottom-right) by varying spline order m and roughness penalty order r (PEN_r). Optimal smoothing is obtained by minimising GCV. d_2 normalisation

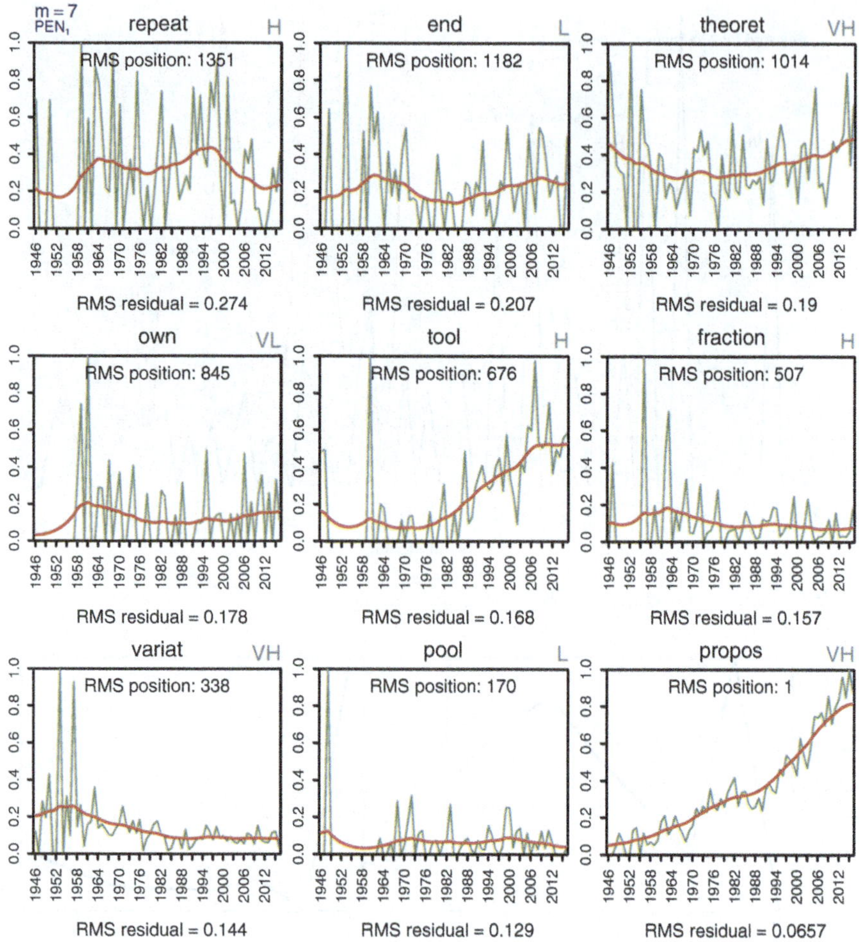

Fig. 9.6 Optimal smoothing: fit a selection of fitted curves ordered according to decreasing root mean square (RMS) residual. Fit of a smoothing spline of order $m = 7$, with PEN_1, to d_2 normalised data

A sample of curves fitted by the optimal smoothing is shown in Fig. 9.4, from the word with highest root mean square (RMS) residual (*repetition*) to the word with lowest RMS residual (*proposal*) (Fig. 9.6).

9.4.4 Curve Clustering

Curves are partitioned by means of the *k*-means algorithm combined with the L_2 metric with cluster number *k* ranging from 2 to 26 and 20 re-runs for each *k*.

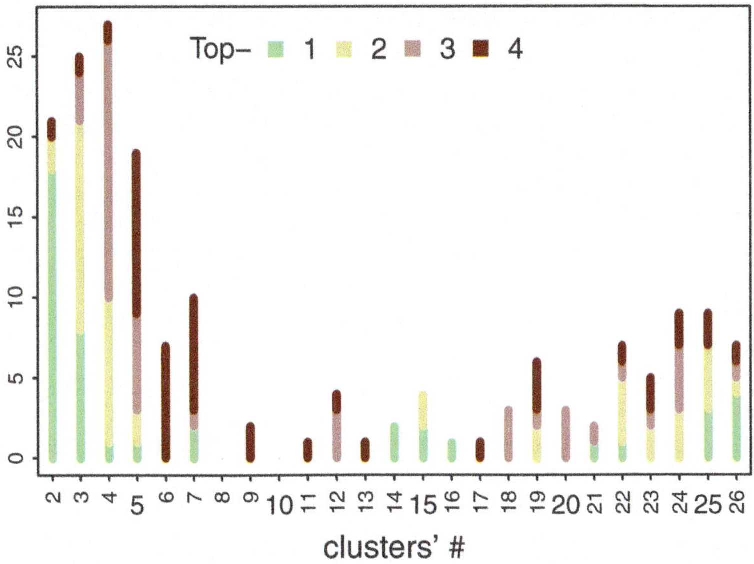

Fig. 9.7 Cluster number selection: frequency of being ranked first (top-1), second (top-2), third (top-3) and fourth (top-4) for each cluster number by pooling rankings from the overall quality criteria. d_2 normalisation

A set of more than 50 quality criteria are then computed in order to identify a set of best candidates for cluster number. Visual representation of the cluster number rating (Fig. 9.7) shows that: (1) partitions into two/three clusters are the best rated; (2) partitions with a cluster number close to the maximum of the considered range have also been frequently selected in the highest positions; (3) in the range of more interesting cluster numbers (neither too low nor too high), the most selected in the top four positions are 4/5 and secondarily 7/15 (in reading the figure, note that bar height corresponds to the cumulated frequency of being in the top four, and colour indicates the position level). This ranking is the output of an R code that essentially mimics a qualitative rating based purely on a graphical inspection.

Discarding the less interesting solutions (1) as well as (2) (which on the one hand may reflect the lack of a defined structure and parsimonious grouping, but on the other may be a failure due to the standard assumption underlying many quality criteria of normally distributed data and hence of compact and convex clusters), the method produces the best partitions corresponding to cluster numbers that emerged at (3) in order to subject them to the scrutiny of experts. In particular, given the set of quality criteria that ranked the cluster number in the top positions and the partitions selected by these criteria, the partition which maximises the average Rand index of agreement with all the other selected partitions is chosen as the best partition for each cluster number.

9.4.5 Substantive Expertise

Here, we illustrate the best partition found with the cluster number ranked first, i.e. $k = 4$. It corresponds to partition 3, out of the 20 replications, as it maximises the average Rand index of agreement with the other partitions (2, 19) selected by multiple criteria (top panel of Fig. 9.8). The graphical output shows the groups, with the cluster mean patterns superimposed, together and individually (Fig. 9.8). Individual clusters are chronologically ordered, from the cluster of words that have tended to disappear to the cluster of emerging words in the period 1946–2016 (A, C, D, B in Fig. 9.8), in order to facilitate the identification of subsequent stages in the knowledge field's evolution.

The multiple Rand index of agreement for the overall partition and for single clusters can provide a measure of the stability of the results. The transcribed words—which are just a subset of group words—are ordered according to both their

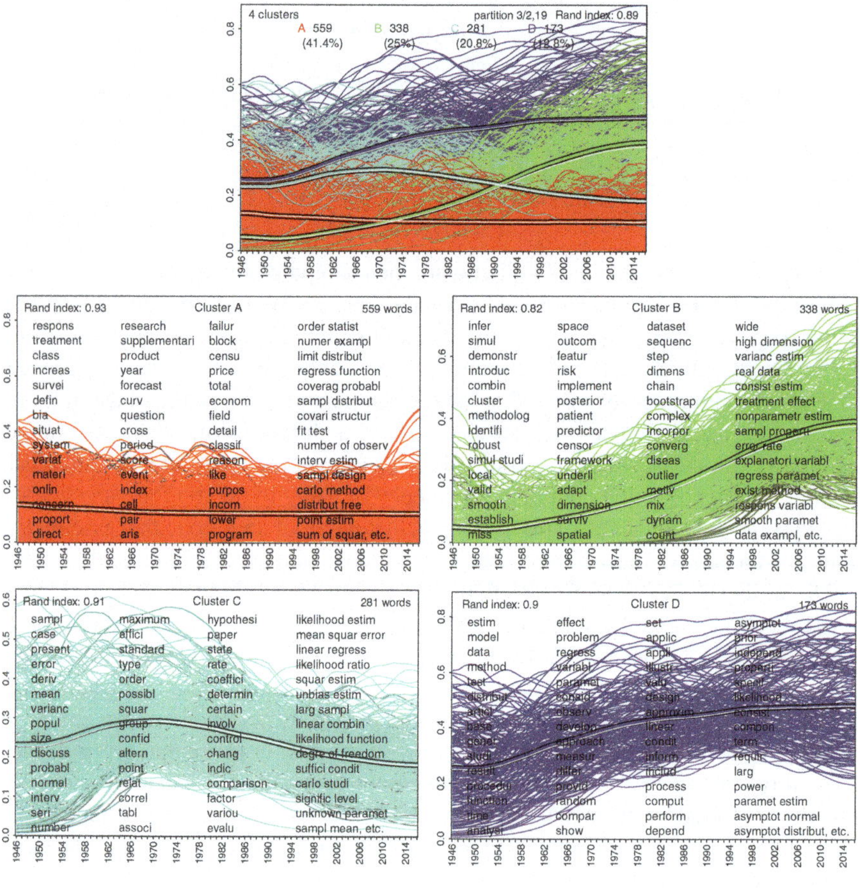

Fig. 9.8 Clustering on d_2 doubly normalised data: all four groups and individual clusters

popularity (from highest to lowest) and their individual multiple Rand index (from highest to lowest). However, a portion of these is dedicated to multiwords whose total frequency is often low or very low (if they are not among the most popular words, they are generally added in the last column of the cluster graph).

The dynamics thus found are then examined and, if considered interesting, are interpreted by subject matter experts. A possible reading of the history of statistics on the basis of the illustrated findings is offered in Chap. 6 where finer partitions—corresponding to the other candidates for cluster number, namely, 5, 7, and 15—are also examined for a more in-depth and detailed interpretation. An interesting nested structure will be found for this particular case of study.

9.5 Concluding Remarks

Normalisation of raw frequencies is critical to reconstruct and compare words' temporal evolution appropriately. The choice of normalisation depends on the interplay between cyclical synchrony and popularity level of words, which underpins the concept of word similarity and ultimately leads to word clustering. This point is discussed in Sect. 9.3.1 and illustrated in Sect. 9.4.

In this study, word trajectories interpreted as FD have been filtered by a basis expansion approach whereby the infinite-dimensional FD are projected onto a low-dimensional space of a set of basis functions. Here, we have chosen B-splines as pre-specified basis functions. A data-driven approach to basis function specification is also possible: functional principal component analysis (FPCA) is a dimension reduction tool that can be used as a method for constructing an optimal orthogonal basis of fixed dimensionality as well. Indeed, functional principal components (FPCs) are often referred to as empirical basis functions. We intend to extend the KBS by providing it with this alternative method of FPC expansion that, among all basis expansions that use K components for a fixed K, explains most of the variation in the FD.

In this study, a two-stage CC via functional basis expansion has been presented. Taking up the finite approximation FPCA approach mentioned above, an alternative can consist of a two-stage CC via FPCA where a k-means algorithm is used on the FPC scores (Peng and Muller 2008). An even more refined method is the FPC subspace projected k-centres functional clustering algorithm (Chiou and Li 2007) whereby cluster centres are identified as subspaces, which account for both the means and the modes of variation differentials between clusters, rather than as cluster means only (like for the k-means algorithm).

Lastly, contrary to two-stage methods, in which filtering is done previously to clustering, we intend to explore strategies performing these two tasks simultaneously (like with model-based techniques). For example, to identify optimal subspaces for clustering and optimal clusters of functions simultaneously, Yamamoto (2012) developed an alternate algorithm which optimises an objective function defined as the sum of the distances between the observations and their projections plus the distances between the projections and the cluster means.

References

Chiou, J. M., & Li, P. L. (2007). Functional clustering and identifying substructures of longitudinal data. *Journal of the Royal Statistical Society: Series B, 69*, 679–699.

Cobo, M., López-Herrera, A., Herrera-Viedma, E., & Herrera, F. (2011). An approach for detecting, quantifying, and visualizing the evolution of a research field: A practical application to the Fuzzy Sets Theory Field. *Journal of Informetrics, 5*(1), 146–166.

Cobo, M., López-Herrera, A., Herrera-Viedma, E., & Herrera, F. (2012). SciMAT: A new science mapping analysis software tool. *Journal of the American Society for Information Science and Technology, 63*(8), 1609–1630.

Desgraupes, B. (2016). clusterCrit: Clustering indices, R package version 1.2.7.

Ding, W., & Chen, C. (2014). Dynamic topic detection and tracking: A comparison of HDP, C-word, and cocitation methods. *Journal of the Association for Information Science and Technology, 65*(10), 2084–2097.

Genolini, C., Alacoque, X., Sentenac, M., & Arnaud, C. (2015). kml and kml3d: R packages to cluster longitudinal data. *Journal of Statistical Software, 65*(4), 1–34.

Griffiths, T. L., & Steyvers, M. (2004). Finding scientific topics. *Proceedings of the National Academy of Sciences, 101*(Suppl 1), 5228–5235.

Jacques, J., & Preda, C. (2014). Functional data clustering: A survey. *Advances in Data Analysis and Classification, 8*(3), 231–255.

Lukas, M. A., de Hoog, F. R., & Anderssen, R. S. (2016). Practical use of robust GCV and modified GCV for spline smoothing. *Computational Statistics, 31*(1), 269–289.

Montero, P., & Vilar, J. (2014). Tsclust: An R package for time series clustering. *Journal of Statistical Software, 62*(1), 1–43.

Peng, J., & Muller, H. G. (2008). Distance-based clustering of sparsely observed stochastic processes, with applications to online auctions. *Annals of Applied Statistics, 2*, 1056–1077.

R Core Team. (2017). *R: A language and environment for statistical computing*. Vienna: R Foundation for Statistical Computing.

Ramsay, J., & Silverman, B. W. (2005). *Functional data analysis (Springer series in statistics)*. New York: Springer.

Rand, W. M. (1971). Objective criteria for the evaluation of clustering methods. *Journal of the American Statistical Association, 66*(336), 846–850.

Shabunina, E., & Pasi, G. (2018). A graph-based approach to ememes identification and tracking in social media streams. *Knowledge-Based Systems, 139*(Suppl C), 108–118.

Trevisani, M., & Tuzzi, A. (2012). Chronological analysis of textual data and curve clustering: preliminary results based on wavelets. In Società Italiana di Statistica (Ed.), *Proceedings of the XLVI Scientific Meeting* (pp. 1–4). Padova: Cleup.

Trevisani, M., & Tuzzi, A. (2013a). Shaping the history of words. In I. Obradovic, E. Kelih, & R. Köhler (Eds.), *Methods and applications of quantitative linguistics: Selected papers of the VIIIth international conference on quantitative linguistics* (pp. 84–95). Belgrad: Academic Mind.

Trevisani, M., & Tuzzi, A. (2013b). Through the JASA's looking-glass, and what we found there. In Proceedings of the *28th International Workshop on Statistical Modelling* (vol. 1, pp. 417–422). Istituto Palermo: Poligrafico Europeo.

Trevisani, M., & Tuzzi, A. (2015). A portrait of JASA: The history of statistics through analysis of keyword counts in an early scientific journal. *Quality and Quantity, 49*(3), 1287–1304.

Trevisani, M., & Tuzzi, A. (2018). Learning the evolution of disciplines from scientific literature: A functional clustering approach to normalized keyword count trajectories. *Knowledge-Based Systems, 146*, 129–141.

Tuzzi, A., & Köhler, R. (2015). Tracing the history of words. In A. Tuzzi, M. Benesová, & J. Macutek (Eds.), *Recent contributions to quantitative linguistics* (pp. 203–214). New York: DeGruyter.

Wagner, S., & Wagner, D. (2007). *Comparing clusterings: an overview*. Universitat Karlsruhe, Fakultat fur Informatik Karlsruhe. Retrieved from https://publikationen.bibliothek.kit.edu/1000011477/812079

Wang, J. L., Chiou, J. M., & Mueller, H. G. (2016). Functional data analysis. *Annual Review of Statistics and Its Application, 3*(1), 257–295.

Yamamoto, M. (2012). Clustering of functional data in a low-dimensional subspace. *Advances in Data Analysis and Classification, 6*, 219–247.

Zhang, Y., Zhang, G., Chen, H., Porter, A. L., Zhu, D., & Lu, J. (2016). Topic analysis and forecasting for science, technology and innovation: Methodology with a case study focusing on big data research. *Technological Forecasting and Social Change, 105*, 179–191.

Zhang, Y., Chen, H., Lu, J., & Zhang, G. (2017). Detecting and predicting the topic change of knowledge-based systems: A topic-based bibliometric analysis from 1991 to 2016. *Knowledge-Based Systems, 133*(Suppl C), 255–268.

Zhou, H., Yu, H., Hu, R., & Hu, J. (2017). A survey on trends of cross-media topic evolution map. *Knowledge-Based Systems, 124*(Suppl C), 164–175.

Chapter 10
Topic Detection: A Statistical Model and a Quali-Quantitative Method

Stefano Sbalchiero

Contents

10.1 Introduction 190
10.2 Latent Dirichlet Allocation (LDA) 191
  10.2.1 Application of LDA 193
10.3 Reinert's Method 199
  10.3.1 Implementation of Reinert's Method 202
10.4 Conclusions 206
References 208

Abstract This chapter aims at comparing and contrasting two approaches for the automatic detection of topics in texts that show interesting similarities and differences. Among the advances that have given a new impetus and vitality to the discipline, the importance to identify topics or, in other words, the procedures that enable us to identify thematic groups within the texts seem to be relevant to meet the scholars' needs and have been developed in different disciplines. Two approaches are compared and contrasted. The first, well-known as Latent Dirichlet Allocation, has been developed as a part of text mining statistical model mainly to classify automatically the texts of large corpora; the second, the Reinert's methods, was developed mainly in the social sciences to managing reliably the content analysis process by bridging the gap between qualitative and quantitative text analysis. Both the procedures proved useful, but in different ways: Latent Dirichlet Allocation enabled us to classify the abstracts automatically under certain topics, while Reinert's method was useful for identifying the internal structure of the abstracts and extracting the macro-topics that characterized them.

Keywords Topic detection · Topic extraction · Latent dirichlet allocation · Reinert's methods · Iramuteq

---

S. Sbalchiero (✉)
University of Padova, Padova, Italy
e-mail: stefano.sbalchiero@unipd.it

A. Tuzzi (ed.), *Tracing the Life Cycle of Ideas in the Humanities and Social Sciences*, Quantitative Methods in the Humanities and Social Sciences,
https://doi.org/10.1007/978-3-319-97064-6_10

10.1 Introduction

The term *text analysis* refers to a whole set of techniques and methods that enable the analysis of textual corpora, or collections of texts, however extensive. In today's world, thanks to an ever greater diffusion of Internet and to technological advances, the amount of textual information with which we all interface every day is enormous. With our social media and social networks, newsgroups, forums and blogs, and information flows, the opportunity to access archives and digital content via the web is constantly posing new challenges regarding the need to monitor the huge quantities of data being generated, and the chance to access and manage the data being consumed every day. Tools have had to be been developed to enable us to orient ourselves, and other users, amidst this unstoppable flow of available resources. Such tools include search engines, which are needed to standardize increasingly advanced procedures useful for automatically analysing, processing and classifying the core information contained in texts. Whenever we input a query in a search engine, and its automated completion procedure generates a number of possible options and presents us with a list of results, we are exploiting applications and algorithms based on a statistical analysis of textual data.

While the digital era has made available enormous quantities of data, technological advances have facilitated the birth and growth of methods for automating the processes of data encoding and analysis. Text analysis is hardly a recent invention (Berelson 1952; Krippendorff 1980; Losito 1993; Tuzzi 2003), but it is only since the second half of the last century that automated approaches to text analysis have made huge progress, thanks to the availability of text corpora in electronic format, and to the development of software for their analysis. For the purposes of the present contribution, among the fundamental advances that have given a new impetus and vitality to the discipline, some are essential to our understanding of the direction taken to develop statistical analyses of textual data designed to identify *topics* or, in other words, the procedures that enable us to identify thematic groups within the texts. These tools are particularly useful for exploring large quantities of documents, extracting essential information, obtaining an overview of the content, and consequently classifying the texts automatically—even in the absence of a previously established classification system. To be more specific, two lines of research have tried to respond to different needs, and have been developed in different settings and disciplines, though they pursue similar goals. On the one hand, we have Latent Dirichlet allocation, developed as a part of text mining methods. On the other, we have Reinert's method which was developed mainly in the psycho-social sciences with a view to managing large quantities of textual matter and bridging the gap between qualitative and quantitative text analysis.

10.2 Latent Dirichlet Allocation (LDA)

The first of the two algorithms presented here is relatively new, and undeniably the topic model that has become the most widespread. Called latent Dirichlet allocation (LDA), it was developed in the world of computer science, machine learning (ML), and text mining in general. Before explaining the model, it is probably worth briefly retracing the steps that led to its development, and the historical and cultural conditions that contributed to its success today. What goes today by the name of text mining (TM) is, without a shadow of doubt, the outcome of a cultural effervescence, and of countless studies in the sphere of language that, as of the 1960s, came together with computer science and technological developments. Strongly featuring cooperation between different areas, this particular discipline has always been defined as a sector with scarcely defined or definable boundaries, in which computational linguistics, statistics, mathematics, some of the social sciences, and computer science and engineering have provided a fertile medium for profitable collaborations. Deriving from data mining (or the set of methods and techniques for extracting information and knowledge from large quantities of data using automated methods), the direction and results of these collaborative efforts have found expression both in basic research, and in applied and industrial research. To give a few examples, suffice it to mention voice recognition, automated translation, information systems, and knowledge management systems in general for the deliberate handling of knowledge using information technologies (Giuliano and La Rocca 2008). TM methods came to light and multiplied in this setting, proving fundamental in efforts to cope with the problems involved in analysing and managing large quantities of freely written or unstructured texts. The goal was to mine these documents to obtain the structured information needed to feed into large-scale databases. These techniques have obviously been hugely successful beyond the industrial and business spheres too, because they can potentially be applied to any type of unstructured text, such as webpages, press agencies, online archives, and so on, and are consequently of interest to many disciplines (Sanger and Feldman 2007). Looking back, albeit briefly, at the steps that led to the success of such applications that we see today, we could say that it all started in the 1960s, with the pioneering lexical-textual studies (Luhn 1959) that focused on indexing (Maron and Kuhns 1960), and the search for documents relating to particular topics. In fact, information retrieval (IR), and information extraction (IE) enabled large corpora to be managed, retrieving documents by means of search queries (IR), or classifying them to make them easier to consult (IE). In the decade that followed, more attention was paid to the opportunity to describe texts with the aid of mathematics, or to introducing innovative information retrieval strategies based on the automatic hierarchical clustering of documents (Jardine and Van Rijsbergen 1971). Later on, some pioneering studies gave birth to computational linguistics, also relying on linguistic meta-information such as electronic dictionaries and word frequency lists (Busa 1974–1980). It was from the 1980s and 1990s onwards, in fact, that an abundance of algorithms came to be developed in the context of artificial intelligence and ML, and then applied to language research (Porter 1980; Berger et al. 1996).

The rest is recent history, and not only has TM developed with a strong interdisciplinary connotation, but its applications have further expanded the body of researchers interested in taking a statistical approach to explore textual data. In a nutshell, the success of TM lies in two trends that have jointly come to the fore. First, there is the diffusion of Internet, social networking applications and technological developments that have made available texts in digital format in a quantity and of a quality that were unthinkable in the 1950s and 1960s. One of the main problems in the past had been the limited opportunity to access textual data in digital form. Second, it has been necessary to reconsider the tools available to researchers and devise new ones to cope with the problems and demands emerging in various areas of research. Recent advances in hardware and software technology have led to a number of unique scenarios in which TM algorithms have been developed and circulated (Aggarwal and Zhai 2012), including the unsupervised learning methods that, by their very nature, require no training data, and the highly successful probabilistic topic modelling algorithms. The enthusiasm for these solutions stems from the fact that such models can potentially be used to extract latent topics from any collection of texts, and then the texts can be organized by applying thematic information extracted directly from the texts in question. Topic models were developed from the intuitions of latent semantic indexing (LSI) (Deerwester et al. 1990) that, though not a probabilistic model, provided the impetus for the development of probabilistic latent semantic analysis (PLSA) (Hofmann 2001). The subsequent extension of this model led to the implementation of Latent Dirichlet Allocation (LDA), a probabilistic generative model first presented in a study published by David Blei, Andrew Ng, and Michael Jordan (2003). Over the years, numerous probabilistic models based on LDA have been proposed, including the author-topic model (Rosen-Zvi et al. 2004), the dynamic topic model (Blei and Lafferty 2006), the pachinko allocation model (Li and McCallum 2006), and the correlated topic models (Blei and Lafferty 2007), to mention just a few. These are all specific models developed to meet the particular needs of certain researchers, but LDA remains the most widespread model in use today, and a milestone in the panorama of topic models.

LDA enables you to identify the topics contained in a corpus automatically on the basis of probabilistic inferences. Inasmuch as every text in a corpus is represented by a set of latent topics, in "statistical natural language processing, one common way of modelling the contributions of different topics to a document is to treat each topic as a probability distribution over words, viewing a document as a probabilistic mixture of these topics" (Griffiths and Steyvers 2004, p. 5228).

The model generates the topics, and the comprehension and interpretation of the generic topic z_i must be deduced from the words comprising it. The topic model uses probability distributions on the words to reconstruct sets of words that form groups of topics, each of which is characterized by the words that are most closely associated (in terms of probability) with a given topic, and that enable the researcher to identify the topic they deal with. Given these premises, we can see why LDA is a generative model: being unable to identify the topics directly, but having the data (words) available, the model reconstructs the latent structure of the corpus a posteriori, starting from the texts and words comprising it. The interaction between the documents being analysed and the latent topics emerges in the probabilistic generative process of LDA as the follow: "for each document in the collection, we generate the words in a two-stage process.

1. Randomly choose a distribution over topics.
2. For each word in the document.

   (a) Randomly choose a topic from the distribution over topics in step #1.
   (b) Randomly choose a word from the corresponding distribution over the vocabulary.

This statistical model reflects the intuition that documents exhibit multiple topics. Each document exhibits the topics in different proportion (step #1); each word in each document is drawn from one of the topics (step #2b), where the selected topic is chosen from the per-document distribution over topics (step #2a)" (Blei 2012a, p. 78). So, if every text consists of a number T of topics ($j = 1\ldots T$), each characterized by certain words, then each of these topics can be represented as a probability distribution on the vocabulary. As a consequence, "if we have T topics, we can write the probability of the ith word in a given document as:

$$P(w_i) = \sum_{j=1}^{T} P(w_i \mid z_i = j) P(z_i = j)$$

where z_i is a latent variable indicating the topic from which the ith word was drawn and $P(w_i \mid z_i = j)$ is the probability of the word w_i under the jth topic. $P(z_i = j)$ gives the probability of choosing a word from topics j in the current document, which will vary across different documents. Intuitively, $P(w|z)$ indicates which words are important to a topic, whereas $P(z)$ is the prevalence of those topics within a document" (Griffiths and Steyvers 2004, p. 5228).

The reader can consult the references for further details (Hall et al. 2008; Schmidt 2012; Blei 2012b), also on the model estimation (Blei and Lafferty 2009), but we can say very briefly here that, being a generative model, LDA reconstructs (or generates) the documents in the corpus by assigning the documents the probabilistic weight of each topic, and then the distribution of the words for each topic, enabling us to identify those with the highest probability level (or, in other words, the most relevant) for a given topic.

10.2.1 Application of LDA

Merely for illustrative purposes, a corpus constructed ad hoc is used both to implement an example of LDA and in the subsequent paragraphs applying the respective analyses. In both cases, the procedures are implemented starting from the same corpus reconstructed after a preprocessing phase. The corpus consists of 1000

Table 10.1 Lexical measures of the corpus of JASA (1000 abstracts randomly extracted)

(*V*) Word-types	6682
(*N*) Word-tokens	143,111
(*V*/*N*)*100 = Type/token ratio	4.7
(*V*1/*V*)*100 = Hapax percentage	32.8

abstracts randomly extracted from the JASA (Journal of the American Statistical Association) database for the years 1946–2016 (see Chap. 6). The corpus of 1000 abstracts (Table 10.1) includes 143,111 word-tokens (N = occurrences) and 6682 word-types (V = different words).

The operations applied in the preprocessing phase were as follows: POS (part-of-speech) tagging; stemming (reducing words to their base form, e.g. *computing* to *compute*); and recognizing Multiword Expressions (MWEs), i.e. sequences of words that gain meaning or change meaning if considered as a whole (see Chap. 8). In all, 370 MWE were included with a frequency ≥ 5. From a lexical viewpoint, the excerpt of the lexical contingency table (Table 10.2) shows that obviously the most common words (*the*, *of*, *and*) are conjunctions and prepositions. The most frequent words are related to the common topics of the abstracts as *model* (1047), *data* (838), *estimation* (753), *method* (659), and *distribution* (545), while the most frequent multiword expressions in this corpus are *simulation study* (65), *likelihood estimation* (38), and *explanatory variable* (35).

Starting from the stemmed corpus reconstructed and including word-types and MWE, LDA was run using the *topicmodels* package (Grün and Hornik 2011) available in the R language for statistical computing and graphics (R Development Core Team 2016), which enables the algorithm proposed by Blei to be implemented in an open-source environment. The corpus had already been reconstructed and stemmed, so the preprocessing phase in R consisted simply in removing stop words (*the*, *if*, *and*, ...), i.e. words that are very frequent in the corpora that are usually omitted in this perspective of analysis. The first question to deal with is the model fitting process because the LDA algorithm demands that the number of topics be specified a priori. It goes without saying that this can be an important and sensitive decision to make, which influences the results. Various ways to deal with this issue have been proposed (Arun et al. 2010; Ponweiser 2012), but the one most often mentioned and appreciated for its simplicity was suggested by Griffiths and Steyvers (2004). It is based on a Bayesian approach that involves computing the log-likelihood for all the possible models in a given interval to identify the maximum value (we normally find the number of topics stabilize within a range of iterations). Applied to the corpus analysed here, this model suggests that the most appropriate number of topics is around 30. Generally speaking, it is necessary to limit the number of topics (whatever the actual figure identified) to make them easier to interpret (Fig. 10.1).

The output of LDA adapted to 30 topics includes a prevalent topic assigned to each abstract, plus the per-document (per-abstract in our case) topic probabilities (Table 10.3).

Given that every document can be represented as a combination of topics presenting different probabilities, an abstract is not necessarily assigned a topic

Table 10.2 Excerpt of contingency table words × years. Occurrences in JASA (1000 abstracts)

Words	Occurrences (corpus)	1946	1947	1948	:	1979	1980	:	2014	2015	2016
the	10,515	130	55	101	:	68	76	:	181	160	202
of	6831	97	63	70	:	41	46	:	99	100	121
and	3889	45	23	34	:	22	31	:	97	70	116
model	1047	0	0	0	:	7	3	:	38	29	48
data	838	11	8	4	:	4	10	:	23	25	25
estim	753	2	3	5	:	1	2	:	14	9	14
method	659	7	2	3	:	6	10	:	12	12	25
distribut	545	1	0	1	:	7	8	:	2	5	5
test	470	4	0	0	:	2	10	:	0	8	3
propos	450	0	0	0	:	2	1	:	19	12	20
effect	431	1	1	1	:	6	0	:	15	10	12
base	426	3	0	1	:	4	2	:	12	8	13
sampl	399	5	12	1	:	4	3	:	6	4	3
studi	371	1	4	1	:	1	0	:	11	15	11
function	360	1	0	1	:	0	3	:	8	6	9
variabl	359	2	0	0	:	4	5	:	11	2	7
measur	337	6	0	1	:	3	0	:	4	3	10
gener	333	8	0	0	:	0	7	:	3	3	10
time	320	3	1	6	:	3	7	:	8	8	10
result	316	3	2	2	:	2	4	:	3	8	7
statist	306	4	3	8	:	3	4	:	4	4	5
present	305	3	5	8	:	1	3	:	1	4	4
simul studi	65	0	0	0	:	0	0	:	2	1	5
likelihood estim	38	0	0	0	:	0	0	:	0	1	0
explanatori variabl	35	0	0	0	:	2	0	:	0	0	0
mean squar error	31	0	0	0	:	0	0	:	0	0	0
high dimension	31	0	0	0	:	0	0	:	2	1	1
axiom	1	0	0	0	:	0	0	:	0	0	0
insignific	1	0	0	0	:	0	0	:	0	0	0
cancerigen	1	0	0	0	:	0	0	:	0	0	0

unequivocally. In the example shown in Table 10.1, a high proportion of abstract No. 7 pertains to topic No. 2, but clearly also refers to matters contained in topic No. 7. To understand this situation, we need to analyse the words characterizing the topics with the aid of lists of words with the highest probability level, i.e. the most relevantly arranged in descending order by topic (Table 10.4).

We can thus check whether abstract No. 7 deals with the problem of sampling by survey (topic No. 2), but also contains information referring to the discussion of similar problems in the case of interviews (topic No. 7). In fact, the article in question is entitled "The Problem of Non-Response in Sample Surveys", and the abstract clarifies the reason why this document has been combined with the two topics:

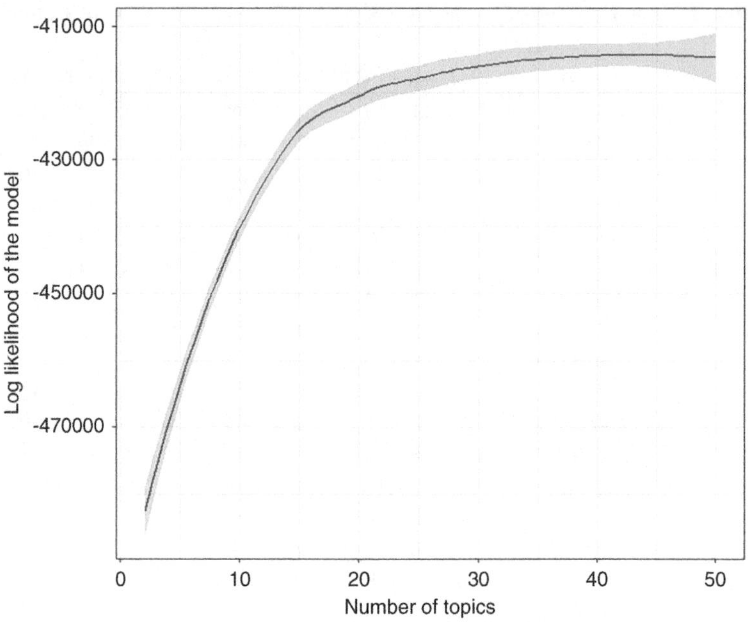

Fig. 10.1 Best number of topics (log-likelihood per number of topics)

Table 10.3 Excerpt of per-document topic probabilities

Id abstract	Topic assignment	Per-abstract topic probabilities					
		Topic 1	*Topic 2*	:	*Topic 7*	:	*Topic 30*
7	Topic 2	0.0002	**0.7708**	:	**(0.0002)**	:	0.0002
129	Topic 1	**0.9941**	0.0002	:	0.0002	:	0.0002
71	Topic 7	0.0004	0.0004	:	0.9883	:	0.0004
899	Topic 7	0.0002	0.0002	:	0.9956	:	0.0002
466	Topic 30	0.0008	0.0008	:	0.0008	:	0.97784

Table 10.4 Word-topic probabilities (Topics nr. 1, 2, 7)

Topic 1	*p*	Topic 2	*p*	Topic 7	*p*
correl	0.06621	sampl	0.01558	interview	0.01990
matrix	0.02862	survei	0.01523	data	0.01472
data	0.02043	estim	0.01517	question	0.01268
variabl	0.01672	bia	0.01463	studi	0.01247
screen	0.01461	method	0.01238	household	0.01151
select	0.01340	popul	0.01125	enumer	0.01105

"The mail questionnaire is used in a number of surveys because of the economies involved. The principal objection to this method of collecting factual information is that it generally involves a large non-response rate, and an unknown bias is involved in any assumption that those responding are representative of the combined total of respondents and non respondents. Personal interviews generally elicit a substantially complete response, but the cost per schedule is, of course, considerably higher than it would be for the mail questionnaire method. The purpose of this paper is to indicate a technique which combines the advantages of both procedures" (Hansen and Hurwitz 1946, p. 517).

On the other hand, abstract No. 129 is incontrovertibly assigned to topic No. 1 (Table 10.1), characterized by the most probable words (Table 10.2) that we find in the abstract:

"Matrix inversion is used in the least squares analysis of data to estimate parameters and their variances and covariances. When the data come from the analysis of variance, analysis of covariance, order statistics, or the fitting of response-surfaces, the matrix to be inverted usually falls into a structured pattern that simplifies its inversion. (…) When the matrix has no special pattern, as in the usual regression problem, the recommended procedure for matrix inversion is the modified square root method" (Greenberg and Sarhan 1959, p. 755).

For the purposes of the present contribution, once we have identified the topics, we can combine the results with other variables pertaining to the corpus. It may be very interesting, for instance (as in our case), to look at the temporal trend of the topics by using a variable not involved in the topic identification process, i.e. the year of publication of the articles (and related abstracts). This possibility was illustrated for the first time by Griffiths and Steyvers (2004): "Analysis at the level of topics provides the opportunity to combine information about the occurrences of a set of semantically related words with cues that come from the content of the remainder of the document, potentially highlighting trends that might be less obvious in analyses that consider only the frequencies of single words" (ivi., pp. 5232–5233).

Basically, the authors demonstrated that the topics can be further analysed by observing them together with the trends seen over the years. They did so by describing two steps (Ponweiser 2012). First, the per-document topic distributions are aggregated using a mean value calculated on all the documents for a given year. We will thus have some topics with a higher mean probability of cropping up at a given point in time. Then, a linear trend analysis is conducted on the topics regressed for the years analysed.[1]

To appreciate how the topics have evolved over time, we can represent them (Fig. 10.2) on a graph showing whether each topic has a positive or negative trend.

Consistent with the idea that topics show different trends through time, this temporal analysis identifies at least three temporal patterns for topics: topics whose trajectory has grown in time and it is increasing over time, topics whose trajectory decreased and topics whose trajectory shows a peak-like behaviour only in a specific

[1] I wish to thank Martin Ponweiser for his support in developing the necessary instructions in R for implementing LDA and obtaining these graphic results.

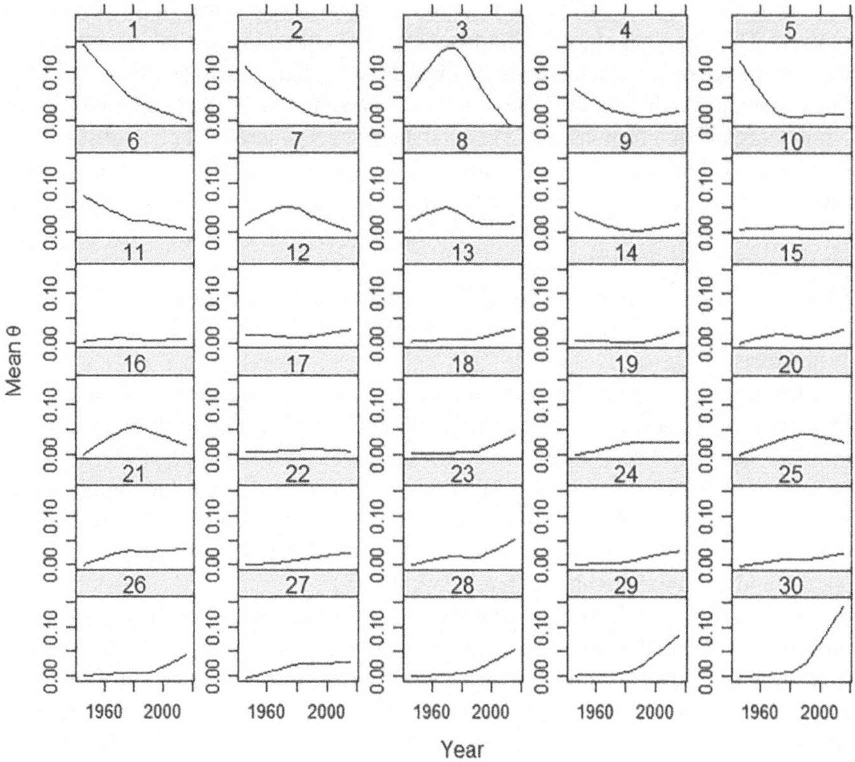

Fig. 10.2 Temporal patterns of 30 topics obtained by linear models (30 panels arranged by slope)

Table 10.5 Number of topics with significant increasing and decreasing trends

	$p \leq 0.05$	$p \leq 0.01$	$p \leq 0.001$	$p \leq 0.0001$
Positive trend	11	8	7	4
Negative trend	6	5	5	5

interval of years. To focus on major increasing or decreasing topics, i.e. on topics that show positive and negative slopes at different levels of p-value, it can be interesting to see how many topics have significantly rising or descending trends (Table 10.5).

The results enable us to restrict the field by analysing the topics that rose or fell considerably in popularity during the years assessed, by means of the significance level of the linear trend test statistic. Considering only those with a statistically significant upward or downward linear trend (p-level ≤ 0.0001), we can identify the "hot and cold topics" (Griffiths and Steyvers 2004) by combining the topic's number with its linear trend (Fig. 10.3).

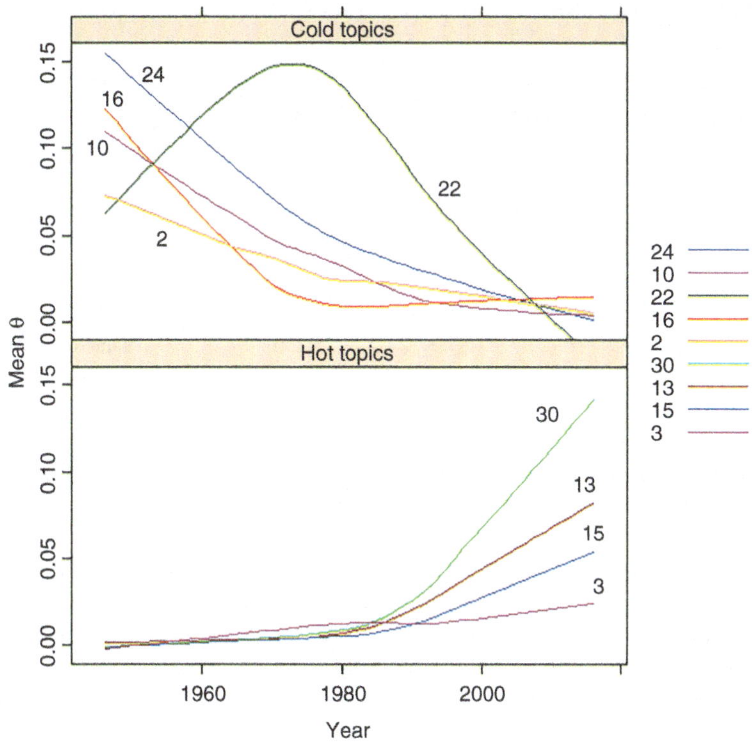

Fig. 10.3 Hot and cold topics (p-level 0.0001)

Looking at the graph with the most probable terms of the five coldest and the four hottest topics (Table 10.6), it is easy to interpret the topics insofar as the most probable words for each topic are connected together, i.e. they refer to the same matter.

Our example is merely for illustrative purposes, so—without going into detail about the content of the topics—suffice it here to say that what we have described is just the tip of the iceberg of all the results produced by the model, which can be used to implement further and more in-depth analyses.

10.3 Reinert's Method

The second approach to topic detection proposed here belongs to the world of tools and automated methods for content analysis. It is a specification that must take into account both the historical evolution of the "classic" and "modern" approaches to content analysis (Tuzzi 2003), and the relationships between quality and quantity (Flick 2014), and between reliability and validity. From a historical perspective,

Table 10.6 Top terms by hot and cold topics (decreasing order of word-topic probabilities)

Hot topics			
Topic 3	Topic 13	Topic 15	Topic 30
bayesian	model	treatment	cluster
mont	time	patient	predictor
infer	effect	effect	distribut
posterior	multipl	studi	approach
mixtur model	paramet	random	nonparametr

Cold topics				
Topic 2	Topic 10	Topic 16	Topic 22	Topic 24
sampl	size	censu	test	approxim
survei	obtain	statist	valu	probabl
estim	problem	standard	normal	formula
bia	varianc	area	bound	rate
method	procedur	measur	random	confid limit

content analysis is certainly not new. In a famous essay on content analysis, Klaus Krippendorff (1980) claimed that the first documented example of text analysis dates back to 17th-century Sweden, when it was applied to a collection of religious hymns known as the *Songs of Sion*. A more recent work, and one of the best known, is a study published in 1918–1920 by Thomas and Znaniecki on "*The Polish peasant in Europe and America*" (Thomas and Znaniecki 1958). This study was conducted on 754 letters exchanged between Polish emigres in the United States and their parents left behind in Poland. From a methodological standpoint, it is an interesting attempt to interpret the letters and classify them manually according to a content-based typology. From the 1920s onwards, mass culture and the diffusion of printed matter motivated quantitative analyses based essentially on article size, column space, and topic frequency (Amaturo 1993). During those years, content analysis received a great contribution from Lasswell's publication "*Propaganda Technique in the World War*" (Lasswell 1927). The author criticized the methodological weaknesses of the research in circulation and coined the name "content analysis" to indicate a type of research that proposed to study the content of propagandistic messages. Lasswell's work is important because applying his content analysis to articles published by the periodical *The Galileian* in a study conducted in 1942 led to William Peley (who was responsible for the periodical) being condemned for disseminating anti-American propaganda. It was Lasswell again who wrote one of the classics of political language (Lasswell 1949), referring to quantitative semantics, promoting the establishment of content analysis in the 1950s, and embracing a much wider analytical universe. This progress suffered a setback towards the mid-1950s, as neatly explained by Sorokin (1956), who coined the term "quantophrenia" to point a finger at the excessive inflexibility imposed by a method that, in pursuing a presumed objectivity, ended up by losing the more qualitative, semantic context of research. The debate during those years led to a sudden change of direction in the search for new methods that could be not only quantitative (an essential feature needed to reduce the complexity of the data), but also more qualitative (to retain the meaning of the content in its rightful context). From this point of view, the work done by Osgood (1959) is exemplary: he proposed new procedures for contingency analysis, that involved working with frequencies without losing sight of the relationships existing between lexical and syntactic elements. He thus paved the way to analyses designed to identify not only the manifest, but also the latent dimension of the message. In the 1960s, in the wake of the success of content analysis in the United States, technological developments and computers prompted a new wave of interest in such approaches. While the development of software enabled far more data to be managed very quickly, it was the linguists who shifted the attention from the mere quantitative aspect to orientations of a lexical type (Tuzzi 2003). In particular, the theoretical and methodological improvements introduced by Jean P. Benzécri (1973a, b)—considered the father of the French school of *Analyse des Données* (Beaudouin 2016)—contributed to the development of a lexical-textual approach that overcame the limits of analysing frequencies alone, focusing more on the relationships between variables from a multidimensional perspective (Benzécri 1982, 1992). This paved the way in subsequent years (especially within the French school) to an encounter between software development and the chance to manage more complex content analyses, expressing a synthesis between quantitative aspects and qualitative factors, and aiming for a statistical exploitation of textual data (Lebart and Salem 1988; Lebart et al. 1998). It is on this stage that we can set the method developed by Reinert (1983, 1990, 1995), and implemented in the Alceste (*Analyse Lexicale par Contexte d'un Ensemble de Segments de Texte*) software, and more recently made available in the R-based version of Iramuteq (*Interface de R pour les Analyses Multidimensionnelles de Textes et de Questionnaires*) (Ratinaud 2014a), used for the present reconstruction. Before explaining the method, which can be defined as a topic detection procedure, it may be useful to add a few comments to place this analytical tool in context and clarify its implications for research. On the theoretical landscape of content analysis (especially in its automated, computer-based version), statistical and linguistic resources are fundamental to enabling analyses with a strong social vocation too. The modern world of content analysis sees linguists, computer scientists, statisticians, psychologists, political scientists, and sociologists embark on advantageous collaborations to develop new tools and paths of investigation capable of responding to the specific and very diverse needs of research. There should be no need to underscore that a great deal depends on the questions being posed, i.e. on the objectives of a given investigation. Using Reinert's method, the two concepts of reliability (the ability to obtain coherent measurements irrespective of the researcher) and validity (the capacity of tools to actually measure what we want them to measure) have proved fundamental in efforts to conjugate scientific strictness with qualitative detail in automated content analysis. It is no accident that it is presented here as a "method". This enables us to position it correctly among the developments of content analysis, while keeping its distance from certain orientations based on the erroneous belief that content analysis could be seen more as a technique for producing results cleansed of the researcher's discretionality and focusing on the manifest content of the communication, according to Berelson's well-known definition (1952). In modern software developments, efforts have focused instead on seeking a compromise between statistical analysis, the researcher's sensitivity and a thorough consideration of the context with a view to optimizing the information available. Comparing the largely manual and qualitative classical approach with the modern one relying on the quantitative tools typical of textual statistics, Reinert's method could be seen as a third option. It takes the need to combine quantity and quality into account by using valid and effective tools to follow pathways that refer to the modern lexical-textual approach, but pursue knowledge goals and objectives that echo the classical approach.

Let us suppose, for instance, that we need to conduct a content analysis on 50 detailed interviews to identify recurring topics relating to how religious beliefs are expressed in the contemporary world. Applying a classical (essentially manual) content analysis, the search will start by encoding the *N* citations identified, then arrange these citations into different topics. To identify significant topics, we need to establish a system of categories (Bryman and Burgess 1994; Braun and Clarke 2006), that we develop inductively (Strauss and Corbin 1990), by constantly comparing emerging results and new intuitions (Savin-Baden and Major 2013; Creswell 2007). The end result will be a set of categories that we can hopefully pool into a limited number of macro-categories that will enable the researcher to test hypotheses, develop theoretical frames, and so on. If the corpus were composed not of 50, but of hundreds of interviews, the job would be difficult to complete manually. Furthermore, in answers to questions in an interview, a given topic is not immediately distinguishable, isolated from other topics being discussed, or clustered around the core issue of a question. Topics are more likely to appear as clues disseminated among several answers (Sbalchiero and Tuzzi 2016). An automated classification of homogeneous portions of text contained in the answers can be useful for extracting topics, or what Reinert (1993) calls "lexical worlds" or classes.

Reinert's (only partly supervised) procedure has the advantage of identifying the lexical worlds contained in any corpus and thus enabling an in-depth investigation on questions and areas of interest, reducing the biases that could develop in the case of a merely qualitative encoding, especially when it is done by different encoders.

10.3.1 Implementation of Reinert's Method

The main goal of Reinert's method is to analyse the organization within a corpus by considering co-occurrences of words as they appear in portions of text, and thereby identify lexical worlds, or semantic classes (Ratinaud and Marchand 2012, 2015; Smyrnaios and Ratinaud 2017). A semantic class (Reinert 1993) is characterized by a specific vocabulary of words associated with one another that form a concrete and observable manifestation of "*topoi*", or conventional themes, and can thus serve as a latent variable (Reinert 1998, pp. 292–293). From a theoretical standpoint, therefore, it is not just a matter of studying occurrences and co-occurrences of words in a text, but of understanding their relationships in the discursive context that, taking this approach, cannot be attributed to chance. Speakers and writers express *topoi* that can only be seen from the lexical tracks they leave when they communicate. When these tracks are repeated more frequently in the discourse, they can be organized into semantic classes (or lexical worlds) that are rich in meaning. The main goal of the algorithm is to shed light on the content and organization of the discourse by separating the lexical worlds from one another, and thus contribute to identifying and constructing vocabularies of co-occurring words that constitute the specific vocabulary of each semantic class.

In operational terms, the procedure involves several successive stages, implemented in the present reconstruction with the Iramuteq software (Ratinaud 2014a). The reference corpus is the same sample of 1000 abstracts drawn from the JASA corpus described in the previous paragraph (and consequently represented by means of stems). The next step consists in the automated identification of minimal units called elementary context units (ECU), i.e. portions of text that may coincide with a phrase, part of a statement, or a paragraph, which are selected on the basis of two empirical criteria: the length of the ECU, in terms of the maximum number of words they can contain (e.g. 40 words); and punctuation marks. These two criteria enable us to divide the corpus into portions of text of similar length. Applied to the corpus analysed, this led to the identification of 3646 ECU containing a mean 39 words each (Table 10.7).

Then the algorithm identifies co-occurrences in each ECU by constructing a contingency matrix of *words x units*. This matrix is organized as a logical table with a repetitive encoding (where "0" = absence, and "1" = presence of a given word form in the portion of text), and this provides the basis for analysing similarities between ECUs, which are summarized by means of a descending hierarchical cluster analysis. The clustering procedure hierarchically identifies the factors (clusters) that best represent a lexical world from the distance of the χ^2 between the classes (Reinert 1983). The result is a dendrogram constructed in successive steps: two classes are created at the start of the analysis, each of which groups together ECUs that reflect a similar lexical content, but differ the most from one another (in other words, an effort is made to reduce the words in common to a minimum). Then we proceed with further distributions of the ECUs by class until they are homogeneous enough to make their further disaggregation impossible (Table 10.8).

The outcome of the classification is a set of classes, each representing a lexical world, i.e. a set of ECUs that include words relevant to the class. From this, we deduce that ECUs are all the more similar, the more words they have in common.

Table 10.7 Corpus and ECU

Number of texts	1000
Mean of occurrences by text	143.1
Elementary context units (ECU)	3646
Mean of forms by ECU	39.3

Table 10.8 Example of table (*words x units*) and dendrogram

	Word 1	Word 2	Word 3	Word 4	Word 5	
ECU 6	1	0	1	0	0	Class 1
ECU 1	1	0	1	0	1	
ECU 3	0	1	0	1	1	Class 2
ECU 5	1	1	0	1	1	
ECU 4	0	1	0	1	1	Class 3
ECU 2	0	1	0	1	1	

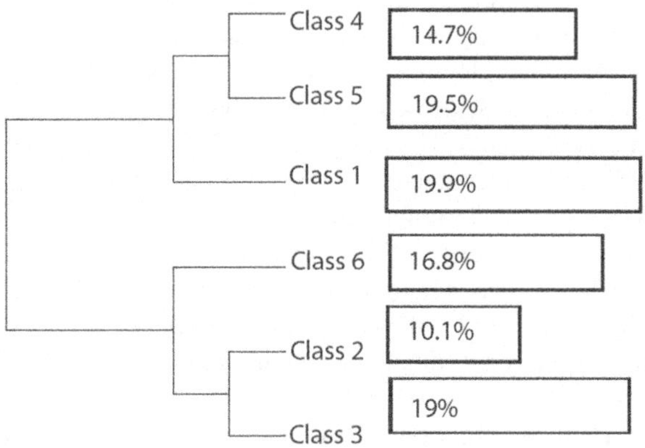

Fig. 10.4 Dendrogram, 6 classes (% of ECU for class)

The list of the most meaningful words, that best represent a lexical world, is identified by associating the χ^2 between words and classes.

In the case presented here, this procedure identifies six classes (or lexical worlds) that account for 81.5% of the ECUs (Fig. 10.4).

In a first step, the results reveal two main clusters, one comprising classes 3, 2, and 6, the second combining classes 1, 5, and 4. In a second step, we can identify another two classes for each main cluster (classes 1 and 6), and the subsequent analysis is run on these last two classes. As we can see, the various steps in the clustering procedure gradually identify somatic groups that have a similar content, and that can be analysed by looking at the word forms significantly associated with each class (Table 10.9).

Finally, the results produced by the algorithm can be used to assess the classes' grade of association with the modalities of other variables, such as year of publication.[2] One of the classic tools for graphically representing this type of result is correspondence analysis, applied to the matrix of a *words x classes* contingency table. But we can also examine the chronological dimension of the lexical worlds using other types of graphical representation (Ratinaud 2014b), such as the proportion of classes by year, and the intensity of classes by year.

[2] Special thanks go to Pierre Ratinaud for his support in developing the R instructions needed to recall the results produced by Iramuteq and construct the graphs from a chronological perspective.

Table 10.9 Vocabulary of classes. Words ordered by decreasing values of association (χ^2)

Class 3	χ^2	Class 2	χ^2	Class 6	χ^2
gene	187.35	patient	507.481	survei	181.754
datum	95.376	treatment	456.508	econom	154.298
network	83.106	clinic	310.868	censu	154.277
biolog	69.297	trial	250.883	employ	143.157
pathwai	64.337	placebo	215.739	period	124.305
tempor	53.767	complianc	188.578	incom	116.824
site	51.417	causal	163.355	labour	114.643
:	:	:	:	:	:
genom	37.136	treatment_effect	106.285	household	96.892
gene_express	36.568	therapi	98.482	industri	91.035

Class 1	χ^2	Class 5	χ^2	Class 4	χ^2
sampl	253.609	regress	206.603	prior	113.497
interv	134.013	estim	149.65	dimension	105.728
popul	117.431	method	124.805	function	93.7
size	96.304	propos	114.543	curv	75.603
confid	88.096	squar	99.383	hierarch	70.604
exact	84.868	invers	90.913	distanc	70.46
varianc	81.397	model	74.998	gaussian	62.477
:	:	:	:	:	:
numer_exampl	34.247	simul_studi	46.763	high_dimension	45.272
sampl_of_size	30.867	real_data	41.353	predictor	42.827

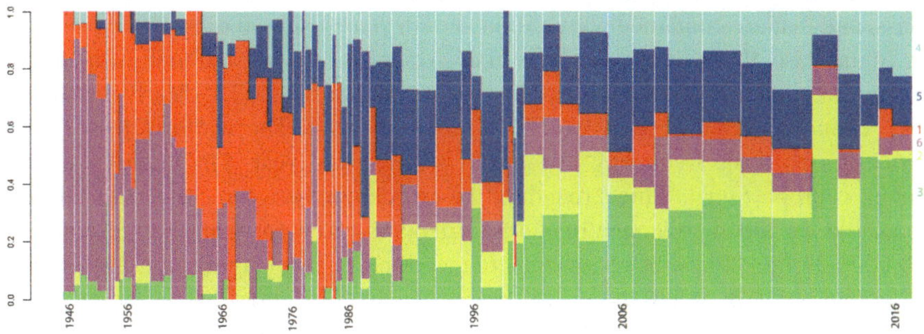

Fig. 10.5 Proportion of classes by year

On the one hand, insofar as a given year can be represented as the sum of all the ECUs extracted from texts published in that year, we can reproduce the proportion of each semantic class from the sum of the ECUs aggregated by class in each year (Fig. 10.5).

The width of the bars is proportional to the number of ECUs in a given year, and their height represents all the ECUs in that year, so we can see the distribution of the classes at a given time. From a chronological standpoint, this representation clearly

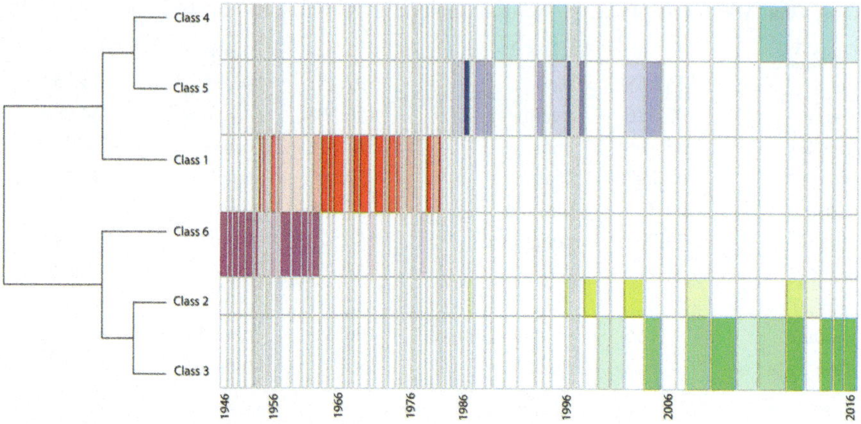

Fig. 10.6 Intensity of classes by year

reveals the trend of the topics over the years: while classes 1 and 6 were typical of the earlier period, class 3 mainly characterizes the years elapsing between the 1990s and the present day; a rising trend can be seen for all the other classes.

On the other hand, we can glean further information from the contributions of χ^2 expressed by the modalities of the "year of publication" variable. If a topic is discussed more during a given period, then the positive differences and the threshold for the significance of χ^2 will indicate an association between year and semantic class given by the relationship between the words most associated with a given class, and that are contained above all in certain years (Fig. 10.6).

In this case, the height of the lines of each class is proportional to the dimension of the class in terms of the number of ECUs it contains. Class 2 is clearly the smallest, accounting for 10.1% of the ECUs (see Fig. 10.4). The width of the cells is proportional to the frequency of the ECUs in a given year. The intensity of the colour is proportional to the strength of the association between class and year: the threshold for the significance of χ^2 has a p-value ≥ 0.05 for the white boxes, and <0.0001 for the darkest boxes, which are therefore the most significant. The shades of colour vary between these two extremes, and this graph thus succeeds in representing what we could call the intensity of the topics over time.

10.4 Conclusions

One of the aims of the present contribution was to assess the performance of two different approaches to topic analysis. The first was developed for the purpose of classifying large quantities of unstructured textual data, the second with knowledge goals more similar to those of qualitative content analysis. Both the procedures proved useful, but in different ways. LDA enabled us to *classify* the abstracts automatically under certain topics, while Reinert's method was useful for identifying the internal structure of the abstracts and extracting the macro-topics that characterized them. Both procedures also enabled us to see how the topics identified varied over time, each generating results that it would have taken hours and hours to obtain manually from the corpus.

One of the problems with both approaches concerns the choice of how many topics to consider. This parameter is very important because the validity of the results depends on the capacity of the model to identify an adequate number of topics. Intuitively, choosing an excessively small number could generate topics that are too broad and heterogeneous, while an excessively large number will give rise to minimal topics that are too specific; either way, they will be difficult to interpret. Neither of the two methods described here is therefore without some degree of discretionality, but this should not necessarily be seen as a limitation. Although there is no unequivocal empirical rule for establishing the parameters that enable the number of topics, k (LDA), and the number of final classes in the clustering procedure implemented by Iramuteq to be increased, researchers can proceed in stages. They can decide whether or not to increase their number depending on the demands of their research, considering the dimensions of the corpus, the number of portions of text that the algorithm succeeds in classifying, the feasibility of interpreting the topics/classes, and so on. If, in a first exploratory analysis, we find topics that include keywords from different sub-topics, we can usually increase these parameters. Alternatively, if we want to analyse a particular topic in more depth, we can create a sub-corpus containing only the documents associated with the topic of interest, or comprising only the ECUs in a given class, on which the analysis can then be repeated. Or again, the outcome of a topic detection process obtained with LDA could be used to further analyse certain topics identified using Reinert's method, given the availability of texts classified under topics: the results would focus specifically on a thematic area, which would be further divided into lexical worlds. Simply put, the possibilities are countless, but it is still a good idea to explain the choices we make during the course of our research, as this is the only way to ensure that the two tools presented here can guarantee reproducible results.

Finally, it is always essential to decide which is the most appropriate approach case by case, depending on the goals being pursued. The automatic analysis of unstructured texts should not be seen as an alternative to the more traditional quantitative approaches, but rather as a means to integrate them by exploiting the availability of a vast range of software and statistical tools that are constantly opening up new opportunities. With developments in the statistical analysis of texts, and the ever-increasing availability of large collections of digitalized empirical material that considerably expand the opportunities for research, the two methods described here can serve as an advantageous meeting place for researchers from different fields. That said, even if the material used in the research is particularly suitable for statistical analysis, the knowledge and expertise needed to interpret the results remain strictly for experts in the sector for the time being.

References

Aggarwal, C. C., & Zhai, C. (Eds.). (2012). *Mining text data*. New York: Springer.

Amaturo, E. (1993). *Messaggio, simbolo, comunicazione*. Roma: NIS.

Arun, R., Suresh, V., Veni Madhavan, C. E., & Narasimha Murthy, M. N. (2010). On finding the natural number of topics with latent dirichlet allocation: Some observations. In M. J. Zaki, J. X. Yu, B. Ravindran, & V. Pudi (Eds.), *Advances in knowledge discovery and data mining* (pp. 391–402). Berlin: Springer.

Beaudouin, V. (2016). Statistical analysis of textual data: Benzécri and the French School of Data Analysis. *Glottometrics, 33*, 56–72.

Benzécri, J.-P. (1973a). *L'analyse des données. 1 La taxinomie*. Paris: Bordas.

Benzécri, J.-P. (1973b). *L'analyse des données. 2 L'analyse des correspondances*. Paris: Bordas.

Benzécri, J.-P. (1982). *Histoire et préhistoire de l'analyse des données*. Paris: Dunod.

Benzécri, J.-P. (1992). *Correspondence analysis handbook*. New York: Marcel Dekker, Inc.

Berelson, B. (1952). *Content analysis in communication research*. Glencoe: The Free Press.

Berger, A. L., Della Pietra, S. A., & Della Pietra, V. J. (1996). A maximum entropy approach to natural language processing. *Computational Linguistics, 22*(1), 39–71.

Blei, D. M. (2012a). Probabilistic topic models. *Communications of the ACM, 55*(4), 77–84.

Blei, D. M. (2012b). Topic modeling and digital humanities. Journal of Digital Humanities, 2(1). http://journalofdigitalhumanities.org/2-1/topic-modeling-and-digital-humanities-by-david-m-blei/

Blei, D. M, & Lafferty, J. D. (2006). Dynamic topic models. In *Proceedings of the 23rd International Conference on Machine Learning* (pp. 113–120).

Blei, D. M., & Lafferty, J. D. (2007). A correlated topic model of science. *Statistics, 1*(1), 17–35.

Blei, D. M., & Lafferty, J. D. (2009). Topic models. In A. Sahami & M. Srivastava (Eds.), *Text mining: Theory and applications* (pp. 71–93). New York: Taylor and Francis.

Blei, D. M., Ng, A. Y., & Jordan, M. (2003). Latent Dirichlet Allocation. *The Journal of Machine Learning Research, 3*, 993–1022.

Braun, V., & Clarke, V. (2006). Using thematic analysis in psychology. *Qualitative Research in Psychology, 3*(2), 77–101.

Bryman, A., & Burgess, R. G. (1994). *Analyzing qualitative data*. London: Routledge.

Busa, R. (1974-1980). *Index Thomisticus Sancti Thomae Aquinatis Operum Omnium Indices ed concordantiae*. Stoccarda: Frommann Holzboog.

Creswell, J. W. (2007). *Qualitative inquiry & research design: Choosing among five approaches*. Thousand Oaks, CA: Sage.

Deerwester, S., Dumais, S. T., Furnas, G. W., Landauer, T. K., & Harshman, R. (1990). Indexing by latent semantic analysis. *Journal of the Association for Information Science and Technology, 41*(6), 391–407.

Flick, U. (2014). *An introduction to qualitative research* (5th ed.). London: Sage.

Giuliano, L., & La Rocca, G. (2008). *L'analisi automatica e semi-automatica dei dati testuali. Software e istruzioni per l'uso*. Milano: Led edizioni.

Greenberg, B. G., & Sarhan, A. E. (1959). Matrix inversion, its interest and application in analysis of data. *Journal of the American Statistical Association, 54*(288), 755–766.

Griffiths, T., & Steyvers, M. (2004). Finding scientific topics. *Proceedings of the National Academy of Sciences of the United States of America (PNAS), 101*(Supplement 1), 5228–5235.

Grün, B., & Hornik, K. (2011). Topicmodels: An R package for fitting topic model. *Journal of Statistical Software, 40*(13), 1–30.

Hall, D., Jurafsky, D., & Manning, C. D. (2008). Studying the history of ideas using topic models. In *EMNLP '08 Proceedings of the Conference on Empirical Methods in Natural Language Processing* (pp. 363–371).

Hansen, M. H., & Hurwitz, W. N. (1946). The Problem of Non-Response in Sample Surveys. *Journal of the American Statistical Association, 41*(236), 517–529.

Hofmann, T. (2001). Unsupervised learning by probabilistic latent semantic analysis. *Machine Learning, 42*(1–2), 177–196.

Jardine, N., & Van Rijsbergen, C. J. (1971). The use of hierarchical clustering in information retrieval. *Information Storage and Retrieval, 7*, 217–240.

Krippendorff, K. (1980). *Content analysis. An introduction to its methodology*. London: Sage.

Lasswell, H. D. (1927). *Propaganda technique in the world war*. New York: Alfred A. Knopf.

Lasswell, H. D. (1949). *The language of politics: Studies in quantitative semantics*. New York: George Stewart.

Lebart, L., & Salem, A. (1988). *Analyse statistique des données textuelles: Questions ouvertes et lexicometrie*. Paris: Dunod.

Lebart, L., Salem, A., & Berry, L. (1998). *Exploring textual data*. Boston: Kluwer Academic Publication.

Li, W., & McCallum, A. (2006). Pachinko allocation: DAG-structured mixture models of topic correlations. In *Proceedings of the 23rd International Conference on Machine Learning* (pp. 577–584).

Losito, G. (1993). *L'analisi del contenuto nella ricerca sociale*. Milano: Franco Angeli.

Luhn, H. (1959). Auto-encoding of documents for information retrieval systems. In M. Boaz (Ed.), *Modern trends in documentation* (pp. 45–58). London: Pergamon Press.

Maron, M., & Kuhns, J. (1960). On relevance, probabilistic indexing and information retrieval. *Journal of the ACM, 7*, 216–244.

Osgood, C. E. (1959). The representational model and relevant research methods. In I. de Sola Pool (Ed.), *Trends in content analysis* (pp. 33–88). Urbana, IL: University of Illinois Press.

Ponweiser, M. (2012). *Latent Dirichlet Allocation in R*. Vienna University of Business and Economics.

Porter, M. (1980). An algorithm for suffix stripping. *Program, 14*(3), 130–137.

R development core team (2016). *R: A language and environment for statistical computing* [software]. Vienna, Austria: R foundation for statistical computing. Retrieved from http://www.r-project.org

Ratinaud, P. (2014a). *IRaMuTeQ: Interface de R pour les Analyses Multidimensionnelles de Textes et de Questionnaires* [software, Version 0.7 alpha 2]. Retrieved from http://www.iramuteq.org

Ratinaud, P. (2014b). Visualisation chronologique des analyses ALCESTE: application à Twitter avec l'exemple du hashtag #mariagepourtous, In *Actes des 12eme Journées internationales d'Analyse statistique des Données Textuelles* (pp. 553–565), JADT 2014, Paris.

Ratinaud, P., & Marchand, P. (2012). Application de la méthode ALCESTE à de "gros" corpus et stabilité des "mondes lexicaux": analyse du "CableGate" avec IRaMuTeQ. In *Actes des 11eme Journées internationales d'Analyse statistique des Données Textuelles* (pp. 835–844), Liège, Belgique.

Ratinaud, P., & Marchand, P. (2015). Des mondes lexicaux aux représentations sociales. Une première approche des thématiques dans les débats à l'Assemblée nationale (1998-2014). *Mots Les Langages Du Politique, 108*, 57–77.

Reinert, M. (1983). Une methode de classification descendante hierarchique: Application a l'analyse lexicale par contexte. *Les Cahiers de l'Analyse des Données, 8*(2), 187–198.

Reinert, M. (1990). ALCESTE: Une méthodologie d'analyse des données textuelles et une application: Aurélia de Gérard de Nerval. *Bulletin de Méthodologie Sociologique, 26*, 24–54.

Reinert, M. (1993). Les «mondes lexicaux» et leur «logique» à travers l'analyse statistique d'un corpus de récits de cauchemars. *Language et Société, 66*, 5–39.

Reinert, M. (1995). I mondi lessicali di un corpus di 304 racconti di incubi attraverso il metodo «Alceste». In R. Cipriani & S. Bolasco (Eds.), *Ricerca qualitativa e computer* (pp. 202–223). Milano: Franco Angeli.

Reinert, M. (1998). Mondes lexicaux et Topoi dans l'approche Alceste. In E. Mellet & M. Vuillaume (Eds.), *Mots chiffrés et déchiffrés* (pp. 289–303). Paris: Honoré Champion.

Rosen-Zvi, M., Griffiths, T., Steyvers, M., & Smyth, P. (2004). The author-topic model for authors and documents. In *Proceedings of the 20th Conference on Uncertainty in Artificial Intelligence* (pp. 487–494).

Sanger, J., & Feldman, R. (2007). *The text mining handbook: Advanced approaches in analyzing unstructured data*. Cambridge: Cambridge University Press.

Savin-Baden, M., & Major, C. (2013). *Qualitative research: The essential guide to theory and practice*. London and New York: Routledge.

Sbalchiero, S., & Tuzzi, A. (2016). Scientists' spirituality in Scientists' words. Assessing and enriching the results of a qualitative analysis of in-depth interviews by means of quantitative approaches. *Quality and Quantity, 50*(3), 1333–1348.

Schmidt, B. M. (2012). Words alone: Dismantling topic models in the humanities. *Journal of Digital Humanities, 2*(1), 49–65.

Smyrnaios, N., & Ratinaud, P. (2017). The Charlie Hebdo Attacks on Twitter: A comparative analysis of a political controversy in English and French. *Social Media + Society, 3*(1), 1–13.

Sorokin, P. A. (1956). *Fads and Foibles in Modern Sociology and Related Sciences*. Chicago: Henry Regnery.

Strauss, A., & Corbin, J. (1990). *Basics of qualitative research: Grounded theory procedures and techniques*. London: Sage Inc.

Thomas, W. I., & Znaniecki, F. (1958). *The Polish Peasant in Europe and America Volumes I and II*. New York: Dover Publications.

Tuzzi, A. (2003). *L'analisi del contenuto: introduzione ai metodi e alle tecniche di ricerca*. Roma: Carrocci.

Chapter 11
What Have We Learnt? Some Concluding Remarks

Arjuna Tuzzi

Contents

11.1 Challenges Met and Challenges Ahead 212
11.2 The Tension Between Established Techniques and Innovative Methods 213
11.3 The Specialisation of Journals 215
11.4 Finding a Balance Between Qualitative and Quantitative 216

Abstract This volume offers various proposals for learning about the temporal development of different disciplines through a (distant) reading of large corpora of scientific literature. Even though these contributions represent only a first step into partially uncharted territory, the results reveal relevant events, moments and timings that corroborate, enrich or change the current narration of the history of ideas. Disciplines and research objects evolve over time, with the modes of communication of research results and scientific language evolving along with them.

The study of large corpora within a diachronic perspective is an innovative and interesting research branch, as today, the vast quantity of texts available for research influence methodological choices. Qualitative analyses are virtually impossible with large corpora, and even semi-automatic work strategies have become increasingly impractical; however, the selection of texts for analysis and the interpretation of results remain qualitative. The tension between qualitative and quantitative methods should be resolved with new mixed-method approaches and well-founded, data-driven computational tools. In the near future, new approaches will not only change how research is conducted in the humanities and social sciences, but also how research is planned and designed.

Keywords High dimensional data · Interdisciplinary research · Qualitative and quantitative approaches · Words life cycle · Words quality of life

---

A. Tuzzi (✉)
Department of Philosophy, Sociology, Education and Applied Psychology, University of Padova, Padova, Italy
e-mail: arjuna.tuzzi@unipd.it

A. Tuzzi (ed.), *Tracing the Life Cycle of Ideas in the Humanities and Social Sciences*, Quantitative Methods in the Humanities and Social Sciences,
https://doi.org/10.1007/978-3-319-97064-6_11

11.1 Challenges Met and Challenges Ahead

The chapters of this volume are concerned with text analysis problems in a chronological perspective. They offer different proposals for learning of the temporal development of a discipline by (distant) reading the temporal evolution of relevant words, keywords and topics in papers published in scientific journals. With reference to different research fields, the authors achieved interesting representations of the temporal evolution of a large set of subject matters and the results often reveal relevant events, moments and timings in the history of their disciplines that either corroborate, enrich or change the current narration of the history of ideas provided by handbooks.

A demanding but very engaging challenge of this research work has been creating an interdisciplinary group of scholars and letting them work together to achieve new representations and new insights for reading the history of their own disciplines. Teamwork in an interdisciplinary environment is in itself a challenging activity, but in this specific case we also asked scholars to work with quantitative methods that in most of their disciplines are neither known nor established. The very idea of treating "texts as data sources" cannot be taken for granted in fields usually explored through qualitative approaches. The authors of this book faced the challenge with great enthusiasm and showed an exceptional, open-minded attitude to research once they decided to get involved in this unfamiliar research terrain.

It seems obvious that this book can only be considered a first step in this uncharted terrain and much remains to be done because the objectives are very ambitious and may require an entire lifetime of work, indeed, as many lifetimes as there are disciplines involved. Even being able to put together a bare minimum review or "state of the art" has been quite difficult in this interdisciplinary perspective and the result is still far from being what can be considered either satisfying or definitive.

While observing the trajectories drawn over time by occurrences of relevant keywords, the research work also opened up interesting theoretical issues concerning the study of the life cycle of words, keywords and topics. The objective is not only to establish the birth date of subject matters by means of the first occurrence of specific keywords in scientific literature but also to study their temporal evolution and eventual disappearance. In the contributions of this book, an initial concept of the words "quality of life" has grown even if we need more sophisticated theoretical grounds to move a step forward and our research tools need to be further refined.

Even in respect to the problem of studying the life cycles of words, this volume shows the potentials of an idea as well as some very interesting results but, obviously, there are still many unanswered questions. Philologists and language historians have always worked on the problem of discovering the first evidence of a word, and similarly, historians of thought and the epistemology of every discipline are seeking the earliest appearance of scientific ideas, tools and concepts. This type of research is difficult because it involves looking back in time for the first evidence of a specific word in a document or finding documented proof in the research work of a scholar that demonstrate that that concept (or that tool, or that idea) had already been conceived, or was at least present in embryonic form in the mind of its creator. This is a work in progress subject to constant updates made possible through the discovery of new documents or the refinement of available research tools.

If dating is already a never-ending, difficult task in itself, then studying the life cycle of words in a dynamic manner is a process which is extremely more complicated because it involves first establishing a birth date and then finding sufficient data to be able to follow its course of life. In our case, we used occurrences in scientific journals as a proxy to measure the vitality of the words and to follow their evolution over time. Naturally, we are aware that the occurrence of a word in a corpus of scientific literature is just one indicator (out of many other elements), i.e. an imperfect measure of its relevance in the general scientific debate. Surely, studying the life cycle of a concept through the occurrences of words in journal articles is a rather unrefined way to be able to provide clear and definitive indicators but according to our analyses, there are many interesting results.

The trend of studying textual data with repeated observations over time of linguistics features is an innovative and interesting perspective in the context of text mining, not only from the epistemological point of view, but also from a methodological point of view. In fact, the quantitative tools to study the trends of occurrences over time need to be refined and the chronological textual data, especially if it is related to large corpora, put existing statistical tools to a real test in terms of precision, as well as from the computational point of view.

11.2 The Tension Between Established Techniques and Innovative Methods

Since it is always important to be updated, often those who do research chose to use new methods even for solving old problems, but sometimes it is also important to know how to take advantage of old methods in entirely new contexts of application.

The methods for quantitative analysis of textual data have been developed in a very different time from the present one in which the availability of texts in digital format was much more limited than it is now. Until 15–20 years ago, even the evaluation of the size of the texts was based on very different criteria from those we use today. For example, a corpus with more than 100,000 occurrences was considered a large text corpus and with a well-constructed corpus of at least 500,000 occurrences it was possible to constitute the reference base for a specialised lexicon. Today, these dimensions are totally outdated and anachronistic. The wide availability of texts in digital format and the inexhaustible source of textual data found in the web and in social networks has completely changed the reference size for research and, therefore, has required a radical updating of adopted methods.

In this new scenario of large corpora and problems of managing high dimensional data, there are some statistical tools such as correspondence analysis that not only persist and are still widely used, but actually have found new lifeblood in the big data world. Correspondence Analysis (CA) is an excellent example of a well-established technique that has found new life in the world of digital methods for text analysis (see Chap. 1). CA offers a way to achieve Euclidean embedding of different information spaces based on cross-tabulation counts and proves useful in analysing great masses of textual data. It can be exploited in information semantics, and particularly in "big data" settings also as a tool suitable for carrying out latent semantic or principal axes mapping in big data scaling.

The large corpora not only result in obvious computational problems but also affect the methodological choices. For example, all semi-automatic approaches to text analysis—those that require manual intervention of researchers who have always played a very important role in the field of text analysis in the past—are today becoming impractical. With the growing dimensions of corpora, the general trend is to give priority to unsupervised and totally automatic approaches.

It is precisely in the logic of moving towards an automated algorithm, i.e. one which does not require the intervention of a researcher, that we have proposed in this volume a knowledge-based system (KBS) to reconstruct the micro-history of each word and identify words that portray similar temporal patterns. In our view, a KBS is a procedure performed by a computer program which incorporates statistical learning techniques and expertise from the knowledge domains involved. The proposed KBS reconstructs the life cycle of words and, by clustering words with similar life cycles, detects exemplary temporal patterns representing the latent dynamics of word micro-histories (see Chaps. 6 and 9). Human intervention comes into play only when the major dynamics uncovered by the KBS are submitted to subject matter experts for the interpretation of results.

This way of studying the trajectories drawn by the frequencies of words is definitely new and innovative in the analysis of chronological textual data even if, from the point of view of the objectives, it has some elements in common with topic detection methods. Both Latent Dirichelet Allocation (LDA) and Reinert's Method provide clusters of words that should reflect a topic as they appear together (co-occur) and, for example, the temporal development of topics is relevant to find hot-topics and cold-topics (see Chap. 10). Nevertheless, differences with the KBS appear evident in the primary research objective: unveiling topics as sets of related words versus outlining life cycles of individual words. These methods for topic detection produce clusters of words that should reflect a topic as they appear together in documents although the shape of word trajectories within topics is not relevant, i.e. they might pool together a set of words in the same topic that do not share a similar temporal development. On the contrary, KBS leads to clusters of words that share a similar evolution over time but that might belong to different topics, different approaches, or different schools of thought.

In the framework of topic detection methods and topic modelling, LDA-based methods (and derivatives) have emerged in recent years as one of the most advanced tools and, at least in theory, they should quickly replace less sophisticated tools such as Reinert's Method. Since this has not happened, it is worth recalling what the elements are that make a method based on a much simpler tool like Reinert's Method preferable to a sophisticated model-based method like LDA.

A first difference is the environment from which they derive: LDA has been developed mainly in the hard sciences to automatically classify the texts of large corpora with reference to their predominant content (topics) while Reinert's Method has been developed mainly in the social sciences to manage the traditional (and qualitative) content analysis process in a more reliable large-scale setting. From this consideration, it is easy to understand that the scholars who come from an environment accustomed to using qualitative methods prefer a tool that "mimics" the working methods of those who do content analysis with a classic qualitative approach work. Although LDA is able to provide much more detailed information (and in terms of probability) on words, on texts and on analysed topics, Reinert's Method (and software packages like Iramuteq and Alceste) continue to be very successful in the field of humanities and social science.

There is still an open question about the effect that the dimensions of the texts have on the results of all these methods. What has emerged from our experience is that when there are texts of very limited length or very long texts, there will always be constraints both in terms of computational problems as well as in terms of readability of the results which deserve further investigation.

11.3 The Specialisation of Journals

Disciplines evolve over time and the modes of communication of research results and scientific language also evolve along with them. When reading articles published in journals in the late nineteenth century, it's not difficult to see that the language used, as well as the writing style, are different than those used today. In addition, some forms of standardisation such as the presence of abstracts or a list of keywords to describe the main contents in articles have become established as best practice at different times but they generally represent a very recent breakthrough for all disciplines. This standardisation of scientific communication sheds light on the effect of a process of learning and sharing a "special language" that is distinctive for each discipline and characterises recent times.

As for the lexicon, most of the analysed journals show a progressive specialisation over time which has become more pronounced in recent decades. For example, through CA we often observed that the scatter of years in the past shows that the range of topics is broader and the lexicon is less technical, while the research areas have become more limited and the lexicon more technical in recent times, especially in the early years of the twenty-first century. However, not all journals reflect this trend.

Although it is not possible to draw general conclusions because the set of analysed journals and disciplines, even though large, is still limited; it seems like the right time to try to explain these trends and leave the task of verifying their meaningfulness to future research.

When we dealt with the young disciplines of the social sciences (sociology, statistics), in the course of history the language of their journals became considerably more specialised and increasingly standardised as the disciplines became established and accredited as autonomous. When they succeeded in affirming their status as autonomous disciplines, they developed their own methods and acquired their own lexicons. Moreover, although today some of the selected journals are still generalist and considered less specialised than others in the field, these journals have increasingly become less generalist and more focused in their contents as an effect of the appearance of other journals in related disciplines, sub-disciplines and research fields or pursuant to the choice of the journal to focus on a certain field of research rather than others (Italian linguistics).

The journals of disciplines which were already autonomous at the time of the journal's founding (social psychology) do not show this marked specialisation over time, and neither those with a millenarian tradition (philosophy) reflect this trend. In short, if a specialisation of the lexicon or an increase in technical terms is not apparent over time, it is because the recognition process of the discipline has already been established for some time and the community of scholars has already acquired a shared and consolidated communicative tradition.

11.4 Finding a Balance Between Qualitative and Quantitative

As already clarified (par. 2 in this chapter), text analysis with quantitative methods makes sense especially in the presence of large corpora, and today the vast quantity of texts available for research makes purely qualitative analyses virtually impossible and even semi-automatic work strategies becomes impractical. However, the selection of texts to analyse and the interpretation of results obtained, which are both crucial step for successful research, remain qualitative (see Chap. 7).

The quantitative analysis of textual data has always hovered between qualitative and quantitative because it is a journey that goes from the word to the number (and back). The analysis of textual data represents an ideal opportunity for the integration of quantitative and qualitative methods and one must remember that texts and text analysis play a key role in research for the humanities and social sciences, perhaps a more important role than for the hard sciences. Naturally, the transition from text to textual data always sacrifices a part of the richness of the text but this is offset by the ability to process large amounts of data in a short time and effectively represent them in a concise way.

The distinction between qualitative and quantitative methods should theoretically be surpassed by mixed-method approaches but, at present, in the humanities and social sciences there is still a deep rift that divides those who believe (wrongly) that it is possible to undertake research without worrying about the quantitative aspects and those who believe (equally wrong) that it is possible to do research without consideration of qualitative aspects of the object of study.

Recent developments of digital methods and the exponential growth of digital humanities suggest that the path is already set and in order to seize the full potential and benefits of this revolution, research needs a new generation of researchers and two different types: researchers with a solid qualitative basis in the field of study but with knowledge of quantitative methods such as to be able to correctly read the results and grasp all the possible criticalities, and researchers with a solid quantitative basis but with knowledge of the field of study such as to fully understand the implications of all the qualitative choices made upstream and downstream of quantitative analysis.

And scholars who already do research cannot stand by idly watching a process which is not only changing how research is conducted but also how new research is planned and designed.

Today, it is not possible to be without knowledge (at least basic) of the new technologies that make it possible to mine large text databases because they are becoming part of communication processes in general, and part of processes of organisation, management and transmission of knowledge in particular. In the near future high performance computers, new generation software packages, methods developed in machine learning and text mining fields, and data-driven computational tools will more and more affect the way scholars conduct research in the humanities and social sciences.

The manufacturer's authorised representative in the EU is Springer Nature Customer Service Centre GmbH, Europaplatz 3, 69115 Heidelberg, Germany. If you have any concerns regarding our products, please contact ProductSafety@springernature.com

Printed and bound by CPI Group (UK) Ltd, Croydon, CR0 4YY
28/11/2025
02007720-0004